A FIELD GUIDE TO DINOSAURS

A FIELD GUIDE TO DINOSAURS

Henry Gee and
Luis V. Rey

BARRON'S

CONTENTS

A QUARTO BOOK

First edition for the United States, its territories and
dependencies, and Canada published in 2003 by
Barron's Educational Series, Inc.
All inquiries should be addressed to:
Barron's Educational Series, Inc.
250 Wireless Boulevard
Hauppauge, New York 11788
http://www.barronseduc.com

Copyright © 2003 Quarto Inc.

ISBN 0-7641-5511-3

Library of Congress Catalog Card No.
2001094437

QUAR.DINO

EARLY AND MID-CRETACEOUS PERIOD 72

LATE CRETACEOUS PERIOD 112

Conceived, designed, and produced by
Quarto Publishing plc
The Old Brewery
6 Blundell Street
London N7 9BH

Editor: Paula Regan
Art Editor: Jill Mumford
Designer: Paul Griffin
Illustrator (cladograms): Dave Kemp
Proofreader: Alice Tyler
Indexer: Pamela Ellis

Art Director: Moira Clinch
Publisher: Piers Spence

Manufactured by
Universal Graphics Pte Ltd., Singapore
Printed by
Leefung-Asco Printers Ltd, China

9 8 7 6 5 4 3 2 1

INTRODUCTION

One thing must be made clear from the start: this is a work of fiction. It is not a scientific document, a report of real findings from the fossil record. Nor is it a wildlife documentary, presenting a reconstruction of the world of the dinosaurs as verifiable fact. This book was conceived as an entertainment, imagining what it would be like to see dinosaurs as living, breathing creatures, in much the same way that a naturalist would observe living animals today.

But all good fiction is enriched by factual research, and we have used the latest paleontological findings as jumping-off points for speculations that are (we hope) sufficiently plausible to convey the experience of dinosaurs as flesh and blood creatures. The age of dinosaurs is extraordinarily remote from our own. The 65 million years that have passed since they disappeared from Earth makes them a million times as distant in time as is World War II from our own day. And, as you would expect, relics from the age of dinosaurs are harder to find, in worse condition, and more difficult to interpret than the memorabilia of Omaha Beach or Pearl Harbor. Virtually everything we know about dinosaurs comes from the crushed and fragmentary fossils of bones and teeth, footprints and eggs. Rare is the dinosaur fossil in which soft tissues such as muscle, skin, and gut are preserved, that might offer a glimpse of dinosaurs as living animals rather than collections of bones.

Many of the dinosaurs you will meet in this book were, until quite recently, unknown. We are living at a time of intense dinosaur discovery, unparalleled since the 19th century, when the exploration of North America brought back cartloads of bones as the pioneers ventured west. Only a decade ago, remains of Sinovenator, Scipionyx, Rapetosaurus, Masiakasaurus, and Carcharodontosaurus among others still lay undisturbed in the ground. The past few years have also seen an explosion in knowledge about dinosaur biology driven by key discoveries, perhaps the most startling of which concern feathers.

OF FUR AND FEATHERS

Some of the newly found dinosaurs you will see here, notably Microraptor, Sinornithosaurus, Shuvuuia, and Beipiaosaurus, were clothed in birdlike feathers, or featherlike fibers, or both, finally confirming a century-old suspicion that dinosaurs were close kin to birds. This implies that many more dinosaurs had feathers than is suggested by the fossils themselves. We have used these facts to award ourselves a certain novelistic license, and we have given many dinosaurs not definitively known to have had feathers a few feathery fringes here and there, and imagined dinosaur young ("chicks") as routinely clad in ducklike down. As we know that many birdlike traits are seen in the remains of dinosaurs that existed before the origin of flight—features such as a clavicle

We know that Microraptor (right) had feathers, and we have used this as the basis for speculating that many of the smaller theropods, including Eotyrannus (above), also sported furry or feathery integuments, although no fossil evidence exists to confirm this.

(wishbone), hollow bones in the limbs, and so on—then we think that such extrapolations are sound and plausible. By the same token, the skins of baby sauropods are known to have been scaly or leathery, so we don't clothe our sauropods in duck down.

During the course of this book we have discussed many features of dinosaur life for which there is little, if any, hard evidence. We imagined the courtship of dinosaurs, depicting the noisy displays of males to audiences of discerning females on a kind of primeval dancefloor or "lek." We speculated that one dinosaur, Scipionyx, was an all-girl institution, reproducing by a process called parthenogenesis in which males are superfluous. The lives of all animals revolve around the need to reproduce, for such is the imperative of natural selection. Life in the wild is notoriously nasty, brutish, and short, and few animals die of old age.

If these speculations sound far-fetched, they shouldn't—every "imaginary" trait that we have applied to dinosaurs is seen, somewhere, in the real-life world. Many birds and mammals display to one another in leks and compete noisily for mates. Several modern amphibian and reptile species are parthenogenetic.

A QUESTION OF COLOR

On the issue of dinosaur color we face a total absence of knowledge. This presents an obvious problem for artists, but plausible color schemes can be inferred in the same way that we can make educated guesses about all other features of dinosaur biology—by comparison with living animals. In general, animals adopt patterns of stripes and spots that allow them to blend in with their background; very large animals tend to be less brightly colored than smaller ones; and animals that are toxic often have very bright colors.

We think that many artists unwittingly view dinosaurs as giant mammals, painting them in the relatively drab palette of the Serengeti, with the occasional relief of a few token spots or stripes. This approach plays down the avian kinship of dinosaurs. If, as seems the case, many dinosaurs shared anatomical features and aspects of behavior with birds, we think it makes sense to clothe them in the bright colors seen in many birds. We'd guess that our dinosaurs are livelier and more colorful than some you might have seen illustrated in other works. Yet we maintain that their bold colors are well within the bounds of scientific plausibility. Many birds and reptiles—unlike most mammals—have color vision, so it seems reasonable to presume that dinosaurs would have responded to the colorful shades of their fellows.

This sustained exercise in imagination should not obscure the many amazing characteristics we portray in dinosaurs that are established fact. Apart from the existence of feathers in an increasing number of theropod dinosaurs, other unique phenomena include the porcupinelike quills of Psittacosaurus (though we've

LEFT If, as seems likely, dinosaurs such as Caudipteryx saw the world in color, then we can speculate that, like modern birds (inset), they would have exploited this faculty and developed bold plumage for display or warning.

invented their venomous stings), the gigantic claws of therizinosaurs, and the weird, stocky forelimbs of Shuvuuia. The remarkable size range of the dinosaurs, from the tiniest Microraptor to the enormous Argentinosaurus, causes us to wonder about these animals and the world in which they lived. The very features that have made dinosaurs the subjects of such enduring interest are, for the most part, scientifically documented realities.

IMAGINING REALITY

Reconstructing the lives of dinosaurs from fragmentary remains is one thing. We've also invented a few species to share the world of the dinosaurs, from microbes and parasitic worms, to tiny flies and immense crocodiles. How can we justify the invention of entire species from nothing at all? The answer is disarmingly simple. The process of fossilization is so uncertain that we can only ever know little about the animals and plants that once shared the Earth. A recent estimate suggests that no more than seven percent of all species of primates that ever existed have been found as fossils, so it is fair to assume that this kind of figure can be applied to other animals, too. Even those dinosaurs with which we are familiar are known from only a handful of specimens— in some cases a single one. Following this logic, it is certain that most creatures that ever lived have left no remains at all. This applies particularly to soft-bodied organisms, and especially to parasites. But parasites are ubiquitous in the modern world. Most contemporary animals are infested with parasites and diseases of every kind, some of which are actually

ABOVE It is not hard to imagine Tyrannosaurus benefiting from the attentions of primeval tickbirds.

vital for the well-being of the species. We humans would have a hard time living without the bacteria that inhabit our gut, and yet as far as we know, no such bacteria accompany fossils of, say, Neanderthal Man. We can, however, safely assume that parasites have been with us since the dawn of life. The fact that they have left little or no trace as fossils does not mean they didn't exist. Indeed, to imagine a real-life biology of dinosaurs without parasites would be a slur against authenticity. Where there once was Triceratops, there was a cloud of worms, bacteria, and viruses that profited specifically from its existence, but which has left no trace whatsoever.

In other words, according to the conventions of fiction, the lives and times we have imagined for our dinosaurs have a kind of "reality." We feel that such speculations as we have made are vital to convey the world of dinosaurs as something the reader might experience on an emotional level, if only vicariously, rather than on the pages of a textbook. We want you to see dinosaurs as more than specimens—we want you to hear their grunts, smell their fetid breath, see the iridescent shimmer of their mating plumage. We want you to feel the heat of Jurassic sunshine on the back of your neck and experience the humidity of a Triassic jungle. We want you to be as startled as we were when we kayaked around the bend of a mid-Cretaceous forest river and saw, illuminated by a ray of sunshine penetrating the high canopy, the first flowers ever to bloom on Earth.

DINOSAUR DISCOVERIES

Dinomania didn't start with *Jurassic Park*—it has been with us since dinosaurs were discovered. Dinosaur bones have been unearthed since antiquity, and like fossils in general, were assumed to have been either works of the devil, or the remains of animals (or even giant people) that perished in Noah's Flood. Certainly, there was no clear idea of the antiquity of the Earth, or that there might have been creatures that had wholly disappeared. But in the late eighteenth century the French savant Georges Cuvier (1769–1832) mooted the concept of extinction, opening the way toward an appreciation of Earth's long and vanished past. From then on the bones discovered were not automatically consigned to the realm of mythology or lore.

The status of dinosaurs as a special and separate group of long-extinct animals dates back only to 1842 when the name "Dinosauria" was invented by the great Victorian anatomist, Richard Owen (1804–1892). Owen looked at the fragmentary remains of

ABOVE Georges Cuvier was a pioneer of geology who championed a radical idea—extinction.

three kinds of extinct reptile—Iguanodon, Hylaeosaurus, and Megalosaurus—that had been discovered recently in England, and realized that these creatures were more than the scaled-up reptiles that had been assumed. They were also different from the seagoing ichthyosaurs and plesiosaurs, already known from the Liassic (early Jurassic) rocks accessible on England's South Coast. Owen recognized that these terrestrial reptiles belonged to a wholly different type of animal—reptilian in basic form, but of greater magnificence, having also some of the verve and vigor of mammals and birds, and less of the sluggardly languor of a snake or lizard. He named them "dinosaurs"—the "terrible lizards." After that, many more dinosaurs were discovered.

THE COLLECTORS

The golden age of dinosaur discovery occurred in the United States during the late 19th century—the legendary period when the West was won. Paleontology had its equivalents of Wyatt Earp and Doc Holliday: Two rivals for the crown of the most prolific describer of new dinosaurs emerging from the remote wildernesses of Colorado and Wyoming. These were Edward Drinker Cope of Harvard (1840–1897), a child prodigy of towering self-regard and Othniel Charles Marsh (1831–1899), more of a late starter, but wise enough to exploit the indulgence of a rich uncle, whom he persuaded to set up a dinosaur museum at Yale, with himself as its director. Most of the dinosaurs with which we are familiar resulted from the competition of Cope and Marsh to haul the mightiest bones out of the West. Not that Cope and Marsh got their own hands dirty too often: They hired the best rockhounds they could find, names that have gone down in paleontological legend, such as Charles Sternberg (1850–1943), who tirelessly toiled for Cope. Sternberg's *The Life of a Fossil-Hunter* is an eye-opening tale of paleontology as it was done on the wild frontier, with the ever-present threat of disease and Apache attack.

While these Ivy Leaguers were slugging it out, more competition arrived in the form of the American Museum of Natural History (AMNH) in New York City. Under the enlightened and sometimes eccentric direction of paleontologist Henry Fairfield Osborn (1857–1935), the museum's search for new dinosaurs would raise its eyes to even more distant horizons. For reasons of his own, Osborn was convinced that the human race originated in the remote wilderness of Central Asia, and he sent an expedition there to find out. This was the famous Central Asiatic Expedition led by Roy Chapman Andrews (1884–1960), one of the most famous fossil hunters of all time and the real-life model for film hero Indiana

ABOVE Othniel Charles Marsh was a professor of natural history at Yale and one of the great pioneers of paleontology. He described 19 new genera of dinosaur.

Jones. In the Gobi desert, Andrews and his team found dinosaur eggs and nests, and the remains of creatures such as Protoceratops. The expansion of the Soviet sphere into Mongolia made it difficult for western teams to work there for many decades, and dedicated teams of Soviet and Polish researchers made steady progress. However, when the Soviet Bloc began to fall apart in the late 1980s and early 1990s, the Mongolian government sent a delegation to the AMNH to ask them to continue where Andrews had left off. The AMNH has sent teams to Mongolia every year for more than a decade since and has made many remarkable dinosaur discoveries, such as the weird dinosaur Shuvuuia, and a fossil of a female Oviraptor preserved while sitting on her nest, shielding her eggs against a sandstorm.

A WORLD OF DINOSAURS

The spotlight has lately turned from Mongolia to China, which has yielded a series of spectacular dinosaur fossils, including the long-necked sauropod Mamenchisaurus, giant hadrosaurs such as Charonosaurus, and other remarkable forms. But most attention has been focused on Liaoning Province in the northeast of the country, which has yielded fossils of exceptional quality and abundance. The key feature of the Liaoning fossils is that they are often preserved with soft tissues as well as bones and teeth. Thousands of specimens of the bird Confuciusornis have been unearthed,

many with feathers intact. The bird Jeholornis has been preserved with intact seeds in its gullet. Several kinds of primitive mammal have been preserved complete with their furry coats. But the most headline-grabbing fossils have been of theropod dinosaurs with feathers, or featherlike pelts, such as Caudipteryx, Microraptor, and Beipiaosaurus. Some of these dinosaurs are featured in this book. These discoveries have changed the way we look at dinosaurs, their lives, and their world.

In the course of gathering this new knowledge, paleontologists have not confined their activities to China or Mongolia, exciting though these countries are. Many recent and remarkable discoveries have been made in far-flung regions such as southern South America (Eoraptor, Herrerasaurus, Giganotosaurus, Argentinosaurus), Madagascar (Masiakasaurus, Rapetosaurus, and others), Southeast Asia (Isanosaurus), North Africa (Spinosaurus, Carcharodontosaurus, Suchomimus), Australia (Minmi, Muttaburrasaurus), and even Antarctica (Cryolophosaurus). But sometimes you don't have to travel too far from home to discover

BELOW A scene from the early Cretaceous of Liaoning Province, northeastern China. In the foreground, two curious Sinosauropteryx (right) approach a pair of Psittacosaurus. Behind them the therizinosaurs Beipiaosaurus browse the trees for insects, while to the right two male Cryptovolans display to each other to assert dominance. A pair of perching Confuciusornis observe the scene.

LEFT The process of fossilization appears so arbitrary that it is a wonder it ever occurs at all. The following sequence illustrates how a fossil might be created. A pair of Allosaurus scavenge the carcass of a Stegosaurus, which has succumbed to an infected wound. Over a period of days, successive waves of scavengers visit the carcass, each further denuding the skeleton of flesh.

dinosaurs—the bizarre fish-eating theropod Baryonyx was discovered by a man while out walking his dog in the UK.

THE ODDS AGAINST A FOSSIL

The popularity of dinosaurs should not disguise the fact that they are quite rare. Fossils form when the remains of living creatures become buried in sediments such as sand or mud, and become impregnated with minerals percolating through the ground water. Their bodies—especially hard parts, such as shells or bones—turn, quite literally, to stone. Fossilization most readily happens in the sea, in which carcasses of animals and plants, especially microscopic ones, rain down on the seafloor. Sometimes the seabed consists of more fossils than rocks. The chalk that is so characteristic of the late Cretaceous—from the Niobrara Chalk of Kansas to the famous White Cliffs of Dover in southern England— consists entirely of the remains of microscopic, marine organisms. These accumulated in enormously thick sediments on the ocean bed. Indeed, the Cretaceous period as a whole gets its name from creta, Latin for chalk. Virtually all the fossils in the collection of any amateur rockhound will be of marine organisms—whether trilobites, brachiopods, ammonites, belemnites, or just clams. There may be a fish or two, and sometimes the saucer-sized vertebra of a marine reptile such as an ichthyosaur. But few people are lucky enough to find dinosaur remains at all.

Fossilization on land is a much chancier business. Animals that die on land often do so in the course of being killed and eaten by another animal. Most of their bodies are digested. Occasionally we may get a view of the end-product of this process: A specimen of paleo-poop attributed to Tyrannosaurus rex is rich in crushed bone. Any leftovers that remain uneaten by the killer, or unscavenged by later opportunists, are worked over by cadres of ever-decreasing size, from insects down to bacteria. In almost all cases, the entire body of a dead animal will be recycled and

nothing will be left to posterity. (The main exception to this rule is teeth. Teeth are coated with enamel, the hardest substance produced by living organisms. Most fossils of land vertebrates—in particular reptiles and mammals—are teeth.) To stand a chance of becoming a fossil, the carcass must become buried almost as soon as its owner dies, so it can rest in peace, rather than rot in pieces.

This may happen in a number of ways. On occasions the body will wash into a lake, perhaps carried by a sudden flood, and become deposited in the ooze of the lake bottom. Sometimes, this ooze will be stagnant—free from oxygen, and thus from the

ABOVE A flash flood engulfs the decomposing animal before all soft tissue has been lost to scavengers and bacteria. The torrential rain brings mud, which quickly covers the carcass, preventing further disintegration of the skeleton. Successive layers of sediment entomb the animal, preserving the articulated skeleton with all but the forelimbs in situ.

oxygen-breathing bacteria of decay. In these rare cases, many of the soft tissues are preserved in addition to bone. The most perfect fossils—complete with feathers and traces of their last meals—are found in mudstones and shales that were once the muddy bottoms of stagnant pools. More commonly, a carcass will wash into a river. As it floats along, it decays, swelling like a balloon with the gaseous by-products of the decay bacteria. When the carcass reaches a bend it gets caught in an eddy and drops into a sandbank. Eventually the corpse falls to pieces, scattering bones along the river bed. Sandstones and mudstones filled with miscellaneous fossil bones may have been laid down in such bone traps in river bends. In very rare cases, an animal is buried alive in a mudslide or even a sandstorm, as was the fate of the Oviraptor mentioned earlier. And there are other ways, too—some fossils have been frozen in ice, pickled in natural brines, ensnared in natural asphalt or even immolated, Pompeii-style, in volcanic ash.

What all these events have in common is their rarity. The probability of a given animal becoming preserved as a fossil is unknowably small, and any dinosaurs we find are prizes indeed. This rarity has an important consequence: It means that the diversity we already see in dinosaur shape and form is just a tiny fraction of what must have existed. This realization is already prompting a reappraisal of our image of dinosaurs as invariably enormous, lumbering beasts. It is easier to find large bones than small ones, and large skeletons make for spectacular museum attractions, and powerful hooks for our imaginations. Add to that the attention from the mass media and an incomplete and inaccurate picture of dinosaur life starts to form in the public's mind, from which it is hard to dislodge. The reality is rather different.

The harder scientists look, the more small dinosaurs they find. We feature many of these in this book. The most exciting small dinosaurs are the birds, of course, and their close relatives among the theropods. These highly evolved creatures are the products of a long, dinosaurian history. But we wouldn't be surprised if most dinosaurs under 7 feet (2m) in length were warm-blooded and insulated with a fibrous or feathery coat. This furry warm-bloodedness may turn out to be an extremely ancient feature, found in the common ancestor of dinosaurs and pterosaurs, which—like most pterosaurs and the earliest dinosaurs—was in all likelihood a rather small animal.

DEEP TIME

Fossils are not only very rare; they are vestiges of periods of Earth's history which, because of their enormous length, are inaccessible to us in any meaningful way. As human beings we measure out our lives in days and weeks—a few decades at most—yet paleontologists discuss intervals of millions of years, incomprehensible to us in all but dry and mathematical ways. John McPhee, in his book *Basin and Range*, coined the term "Deep Time" to refer to such immense chasms of time.

The dinosaurs lived in the Mesozoic era, which lasted from around 245 million to 65 million years ago. The Mesozoic is

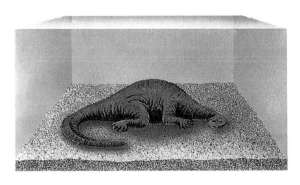

1. A sauropod carcass lies at the bottom of a lake. Sediment begins to settle on the form, giving further protection from erosion.

2. Layers of sediment build up, completely covering the bones. Many years pass, and the lake drains.

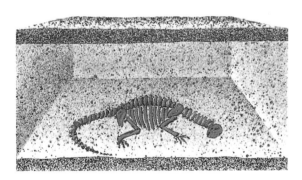

3. The bones, now buried under layers of sediment, are subject to changes in the rock: The bony tissue is replaced, crystal by crystal, by another hard mineral.

4. Over time, the rocks may be tilted by movements in the Earth's crust, exposing the ancient sediments and the fossils they contain.

divided into three shorter (but still very long) intervals, or periods. The first of these, the Triassic, lasted from 245 million to 208 million years ago. The Jurassic period (208–146 million years ago) was the noontide of the dinosaurs, especially the giant sauropods. The subsequent Cretaceous period (146–65 million years ago) saw the greatest diversity of the dinosaurs and ended with their abrupt extinction. The Mesozoic era itself, long though it was, was but a small interval in the history of the Earth. It is thought that the Earth formed around 4,500 million years ago. The first animals and plants big enough to be seen with the naked eye evolved 600 million years ago, by which time almost nine-tenths of Earth's history had already taken place. Between 600 and 500 million years ago there was an evolutionary explosion in which most forms of modern animal life appeared, including the first vertebrates. Animals and plants had made tentative excursions onto land by 400 million years ago, and these included—by

KEY TO DINOSAURS

1 Coelophysis	25 Carnotaurus	45 Tyrannosaurus
2 Eoraptor	26 Baryonyx	46 Giganotosaurus
3 Herrerasaurus	27 Eotyrannus	47 Saltasaurus
4 Liliensternus	28 Hypsilophodon	48 Masiakasaurus
5 Plateosaurus	29 Iguanodon	49 Rapetosaurus
6 Isanosaurus	30 Scipionyx	50 Spinosaurus
7 Allosaurus	31 Aegyptosaurus	51 Deinocheirus
8 Ceratosaurus	32 Carcharodontosaurus	52 Gallimimus
9 Diplodocus	33 Ouranosaurus	53 Oviraptor
10 Ornitholestes	34 Suchomimus	54 Protoceratops
11 Stegosaurus	35 Beipiaosaurus	55 Shuvuuia
12 Archaeopteryx	36 Microraptor	56 Therizinosaurus
13 Compsognathus	37 Psittacosaurus	57 Velociraptor
14 Scelidosaurus	38 Sinovenator	58 Charonosaurus
15 Brachiosaurus	39 Sinornithosaurus	
16 Mamenchisaurus	40 Minmi	
17 Tuojiangosaurus	41 Muttaburrasaurus	
18 Yangchuanosaurus	42 Edmontonia	
19 Cryolophosaurus	43 Pachycephalosaurus	
20 Massospondylus	44 Triceratops	
21 Acrocanthosaurus		
22 Deinonychus		
23 Zuniceratops		
24 Amargasaurus		

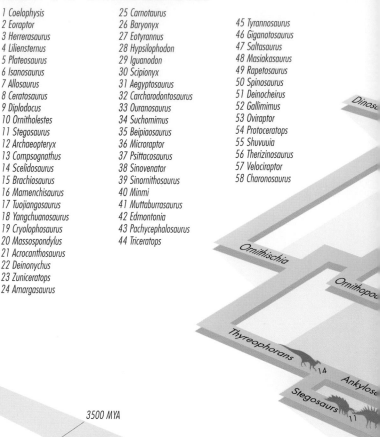

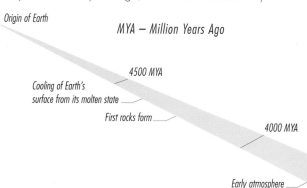

Origin of Earth

MYA – Million Years Ago

4500 MYA

Cooling of Earth's surface from its molten state

First rocks form

4000 MYA

Early atmosphere

3500 MYA

First bacteria appear

3000 MYA

2500

Cyanobacteria oxidizes dissolved iron in oceans

360 million years ago—the first amphibians. After the dinosaurs died out, the role of the giant, ferocious, two-legged carnivore, lately vacated by T. rex, was occupied by flightless birds—oversized relatives of modern cranes. Their reign was mercifully brief, and gave way to a prolific flowering of mammals that continued until relatively recently. And the rest, as they say, is history.

A CLASS OF THEIR OWN

To classify—to impose a sense of order on disordered nature—is a natural human urge, and has been ever since Adam was exhorted to name the beasts in the Garden of Eden. The first classifications of dinosaurs were, or course, pre-evolutionary. Dinosaurs were classified as reptiles, and to begin with, Owen's name "dinosaur" was applied to any large, terrestrial reptile from the Mesozoic. In the 1860s, in the wake of evolution, Darwin's friend Thomas Henry Huxley (1825–1895) noticed how birdlike dinosaurs were. He devised a new reptilian order, the Ornithoscelida, which was divided into two suborders, the Dinosauria (Iguanodon and its relatives, carnivores such as Megalosaurus, and armored dinosaurs such as Scelidosaurus) and the Compsognatha (for Compsognathus and other small, birdlike forms).

In the modern scheme, dinosaurs belong to a large group of reptiles called the archosaurs ("ruling reptiles") which includes crocodiles and a number of extinct forms such as the flighted pterosaurs. Other reptiles, such as turtles, lizards, and snakes, are not classified as archosaurs. The Dinosauria itself is divided into two large groups, the Ornithischia (bird-hipped) and Saurischia (lizard-hipped), distinguished by the shape of the pelvic bones.

The Ornithischia contains a number of herbivorous dinosaurs, including the ornithopods (Iguanodon and its relatives) and their Cretaceous offshoots, the hadrosaurs; ceratopsians such as Triceratops; and the armored dinosaurs or Thyreophora, including stegosaurs, ankylosaurs, and their relatives.

The Saurischia includes the giant sauropods such as Brachiosaurus and Diplodocus, as well as the theropods, that vast panoply of mostly carnivorous dinosaurs from tiny Microraptor to enormous Tyrannosaurus, with all kinds of wonderful creatures in between, ranging from mysterious Deinocheirus to bizarre Therizinosaurus. The theropods include all dinosaurs known to have

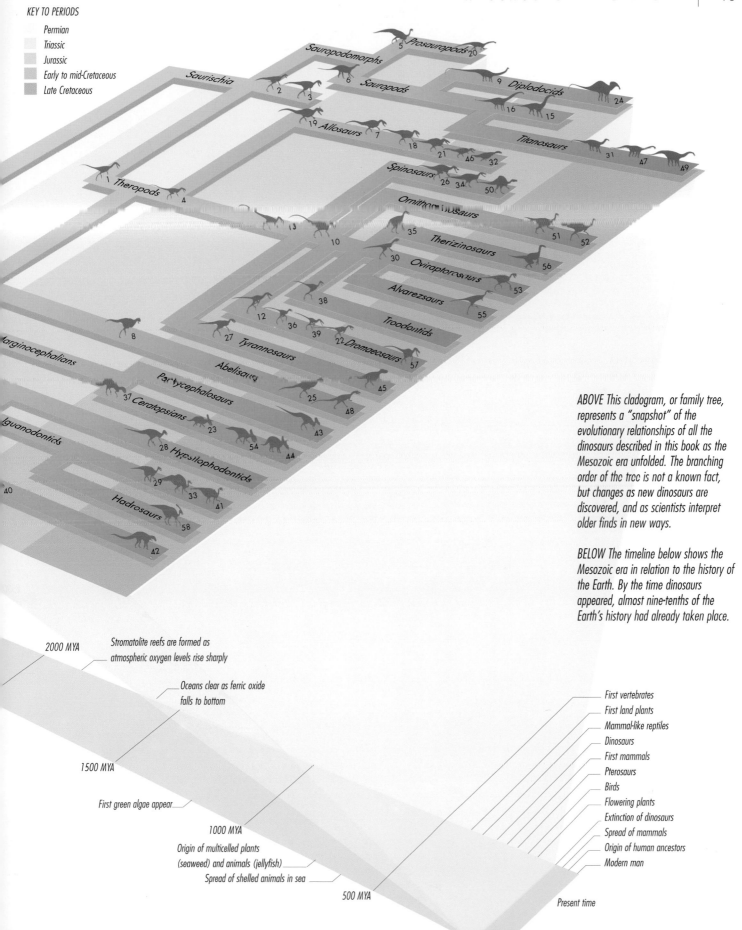

KEY TO PERIODS

Permian
Triassic
Jurassic
Early to mid-Cretaceous
Late Cretaceous

Saurischia
Sauropodomorphs
5 Prosauropods 20
6 Sauropods
9 Diplodocids
2 3
24
16 15
19 Allosaurs 7
Titanosaurs
18 21 46 32
31 47 49
Theropods
1 4
Spinosaurs
26 34 50
Ornithomimosaurs
13
35 51 52
10 Therizinosaurs
30 Oviraptorosaurs 56
Alvarezsaurs 53
38 55
12 Troodontids
8 36 39 22 Dromaeosaurs
27 Tyrannosaurs 51
Marginocephalians Abelisaurs 45
Pachycephalosaurs 25 48
37 Ceratopsians
Iguanodontids 23 43
28 Hypsilophodontids 54 44
29 33 41
40 Hadrosaurs 58
42

ABOVE This cladogram, or family tree, represents a "snapshot" of the evolutionary relationships of all the dinosaurs described in this book as the Mesozoic era unfolded. The branching order of the tree is not a known fact, but changes as new dinosaurs are discovered, and as scientists interpret older finds in new ways.

BELOW The timeline below shows the Mesozoic era in relation to the history of the Earth. By the time dinosaurs appeared, almost nine-tenths of the Earth's history had already taken place.

2000 MYA — Stromatolite reefs are formed as atmospheric oxygen levels rise sharply

Oceans clear as ferric oxide falls to bottom

First vertebrates
First land plants
Mammal-like reptiles
Dinosaurs
First mammals
1500 MYA
Pterosaurs
Birds
First green algae appear
Flowering plants
Extinction of dinosaurs
1000 MYA
Spread of mammals
Origin of multicelled plants (seaweed) and animals (jellyfish)
Origin of human ancestors
Modern man
Spread of shelled animals in sea
500 MYA

Present time

had feathers—including, of course, the birds themselves. Theropods have accounted for the lion's share of the exciting dinosaur discoveries of recent years, so we make no apologies for what might seem like disproportionate coverage.

In this book we have adopted the surprisingly modern convention that no dinosaur is regarded as "ancestral" to any of the others. Prosauropods, for example, are usually seen as more primitive than sauropods, and in many cases lived earlier. But this does not mean that prosauropods were in any sense ancestors of sauropods—at most we can say that they were cousins.

To accommodate Deep Time, paleontologists tend to draw up classifications according to a scheme in which all the animals are regarded as cousins to a greater or lesser degree, irrespective of the period in which they lived, and with no assumptions made about ancestry and descent. It is called cladistics and you can read more about it in Henry Gee's book *In Search of Deep Time*.

DINOSAURS AND BIRDS

The long-running debate concerning the relationship between birds and dinosaurs is less about fossils than our approach to them. Since the time of Huxley, it has been apparent that dinosaurs and birds have much in common. Paleontologists have been happy to think of birds as archosaurs in general, and even as close relatives of particular groups of dinosaurs within the archosaurs. Problems such as the origin of flight in this or that dinosaur group were seen as subsidiary concerns. After all, the anatomy seemed clear enough, and worries about whether these dinosaurs could have flown or not seemed of secondary importance.

More recently, some ornithologists have contested this view, saying that the origin of birds is intimately associated with the origin of flight. In other words, birds started out as small, lizardlike creatures living in trees, adopting progressively more airborne habits, and so gradually turning into birds. This group of scientists contends that dinosaurs could not have been ancestral to birds because dinosaurs were ground-living bipeds, rather than small, tree-living quadrupeds. The earliest known birds (Archaeopteryx) lived millions of years before the theropods held to be the closest relatives of birds. Also, certain details of the anatomy of the hands in birds are so different from the equivalent features in dinosaurs that the former could not have evolved from the latter.

The first objection is easily refuted, for one cannot make or break a scientific theory based on an untestable assertion about what may or may not have happened in the deep past. The second is also easily countered, by reason of the unknowable incompleteness of the fossil record. Archaeopteryx is still the earliest known bird, living at the end of the Jurassic period, 150 million years ago, up to 20 million years before the many birdlike dinosaurs recently recovered from the early Cretaceous of China (145–124 million years ago). But this fact is, in itself, no argument

against a close relationship: All it means is that the origin of birds must be sought well back in the Jurassic. Given the known incompleteness of the fossil record, this comes as no surprise. After all, an archaic kind of fish called the coelacanth was found alive and well in the twentieth century, a full 80 million years after it was thought to have become extinct. This is about as likely as walking down a desert track in Mongolia today and seeing a pack of live, toothy Velociraptors come frolicking toward you.

The third objection—based on the anatomy of the dinosaur and bird hand—is more serious. Like most land vertebrates, reptiles started with a five-fingered hand. The hands of primitive dinosaurs also have five fingers, but in many theropods, including those considered most closely related to birds, this pattern is reduced to three. There is good evidence that the three remaining fingers in these theropods are equivalent to the thumb, index, and middle fingers. In technical parlance, dinosaurs retain digits I, II, and III of the original hand. In modern birds, the hand has become part of the wing and is extremely reduced and modified, yet it still seems to be based on a three-fingered pattern. However, studies on the development of the hand in bird embryos suggest that the fingers of birds correspond to the index, middle, and ring fingers—in

ABOVE This as yet unnamed fossil from the early Cretaceous of Liaoning Province, China, shows clearly the feathery integument on this small theropod, presenting the clearest possible evidence of the relationship between dinosaurs and birds. OPPOSITE Luis Rey's depiction of the "fuzzy raptor" holding its captured prey, the early bird Confuciusornis.

FAR LEFT A small, tree-climbing reptile from the Permian period

LEFT Some Triassic reptiles showed adaptations for running, climbing, and gliding. Some could have been the ancestors of dinosaurs and birds

FROM PRIMITIVE REPTILE TO BIRD...

...in eleven easy steps. One of the many possibilities—purely conjectural—of how birds could have evolved (from an idea by George Olshevsky, as modified by Luis V. Rey).

BELOW A common ancestor of dinosaurs and birds, this Triassic form has insulatory quills, and fringes on the tail and forearms

LEFT Bird-sized common ancestor of Ceratosaurus and birds. Fifth digit is gone and fourth is vestigial

RIGHT Ancestral dromaeosaur. It has three claws, a foldable "wing," and a stiff tail lined with feathers

RIGHT Archaeopteryx, the earliest known bird

BELOW Common ancestor of Allosaurus, advanced theropods, and birds. It has three fingers and protofeathers all over the body

BELOW RIGHT Ichthyornis, an early "modern" bird. The fingers are fused and lack claws, and teeth are still present in the jaw
BELOW LEFT Iberomesornis belonged to a group of flying birds, the enantiornithines, unrelated to modern flying birds

ABOVE A bald eagle—a fully modern bird

LEFT A scene from the early Cretaceous of China: A pair of the perching dinosaur Epidendrosaurus (left) wheedles grubs from bark using their massively extended third digits, while a flock of Microraptor (right) use their four "wings" to leap rapidly through the canopy.

other words, digits II, III, and IV. This poses problems for advocates of a close bird-dinosaur relationship, and this issue remains unresolved. But the embryological work on which this argument is based is subject to a degree of interpretation: It is very hard, when looking at a blob of embryonic cartilage under a microscope, to assert confidently that it represents the germ of this digit or that. It is also possible that digits change their identity during development, so that what starts out looking like digit I in the embryo may end up as digit II in the adult. And, of course, we have no equivalent embryos of dinosaurs.

Nevertheless, paleontologists can point to a large number of features that birds and dinosaurs share, and argue that these outweigh this single point of debate. Some of these features include the presence of a fused clavicle or "wishbone," the tendency for certain bones to be hollow (which in birds accommodate extensions of the lungs), the tendency for certain bones in the legs to fuse, as well as those in the skull to be remodeled in particular ways. Great similarities may also be found in the bones of the wrist and how the wrist would have worked—it is suggested that some dinosaurs folded their hands sideways, in the same way that a bird furls its wings. Paleontologists take the view that this weight of evidence speaks strongly in favor of shared common ancestry between birds and dinosaurs; ornithologists, on the other hand, put it down to convergence—the phenomenon in which the features of otherwise unrelated groups come to resemble one another.

This debate was already well advanced by the late 1990s when Chinese researchers announced the discovery of several kinds of dinosaur with feathers. To a paleontologist, the presence of feathers is just one more item on the long list of similarities between birds and dinosaurs. So much so, that many paleontologists (and artists—including Luis V. Rey) speculated that dinosaur feathers might, one day, be found. To the ornithologists, however, the discovery of feathers in dinosaurs came as a bitter blow. The presence of feathers is seen as iconic: Feathers are seen today only in birds, and all modern birds have feathers. Feathers are the final sign of the evolution of flight in birds. On this reasoning, feathers have come to stand for the indissoluble evolution of birds and flight. However, it follows from none of this that feathers should not have evolved in animals besides birds—perhaps others in the wider group of animals from which birds also evolved. The killer punch is that most dinosaurs with feathers do not appear to have been any more flight-capable than a block of concrete. The evolution of feathers happened before the origin of flight. The ornithologists went to some lengths to discredit these findings, and discover non-dinosaurian fossil reptiles with evidence of feathers, but these efforts have come to nothing. An interesting aside is the idea—advanced several times over the years and most recently promoted by paleontologist and artist Gregory Paul—that many apparently land-living dinosaurs are secondarily flightless, in the manner of, say, modern ostriches or penguins. That is, they

MEET THE DROMAEOSAURS

The closest extinct relatives of birds. From left to right: Bambiraptor, Rahonavis (flying), Sinornithosaurus, Deinonychus, and Velociraptor. The leg of the 23-feet (7-m) long Utahraptor appears in the background

LEFT In the mid Triassic, around 225 million years ago, most of Earth's landmass was fused into one supercontinent called Pangea lying north-south along its axis.
BELOW By the early Jurassic, 180 million years ago, Pangea had begun to break up as Gondwana began to head north.

evolved from flighted ancestors whose remains have yet to be found. Were some dinosaurs really dragons that fell to Earth?

The family tree of dinosaurs we present in this book is based on much recent research and does not pretend to be the Last Word. Our ideas about the interrelationships of dinosaurs change almost weekly as new dinosaurs are discovered, each with its own set of evidence about its place in the wider scheme of things. The most fluid parts of the family tree are to be found among the theropods, currently a particularly active research area, given that all the feathered dinosaurs so far discovered fall into this group. The interrelationships of sauropods is also a topic of current interest, as is the position of some primitive theropod-like dinosaurs such as Herrerasaurus and Eoraptor. Some people think these are primitive theropods, while others regard them as more primitive and generalized dinosaurs, branching from the tree before the division between theropods and sauropods was clearly established.

THE MESOZOIC WORLD

The age of the dinosaurs spanned almost the entire Mesozoic era —183 million years. This interval is bracketed by two abrupt phases of transition, namely the mass extinctions at the end of the Permian and the Cretaceous. The face of the Earth changed markedly during this long interval, by the process of continental drift—a phenomenon so slow as to be invisible to us humans with our short lives, but which, over sufficiently long time scales, dominates life on Earth, and how that life evolves.

Until quite recently, geologists thought that the positions of the continents relative to one another had been fixed since the beginning of time. Although this idea was repeatedly challenged by unusual and disparate findings, these were explained away in the context of the time. Thus the discovery of essentially the same creatures on widely separated continents was explained by the idea of "land bridges" that had once connected them; findings of fossil shells in the rocks of mountain tops were thought to reflect dramatic rises (and falls) in the level of the sea. Some, however, did call this fixity into question, noting, for example, the curious circumstance by which the northeastern coast of South America could be made to fit, jigsaw-puzzle-fashion, into the Gulf of Guinea, off the Atlantic coast of Africa. Again, many fossils found in northern Europe were present in very similar kinds of rocks in North America. Such dispositions could hardly be the result of coincidence. Were North America and Europe, or South America and Africa, once joined up, only to have become separated by

some as yet unknown process? The problem was that no plausible mechanism for continental drift had been advanced.

The breakthrough came with the idea—backed up by evidence from geophysical surveys of the ocean floor—that all parts of the Earth's crust, whether continental or oceanic, formed part of a single, integrated system in which continental drift, and many other circumstances besides, could be explained. The Earth's surface, as it turns out, is divided into distinct regions called "plates," rigid slabs of solid rock that ride atop hotter, more mobile material. The plates are demarcated in the oceans by so-called "mid-ocean ridges"—submerged chains of volcanoes—and by earthquake fault lines and deep scars or "trenches" in the ocean floor. Molten rock wells up from the interior of the Earth through mid-ocean ridges, creating new ocean floor. As more rock is produced, the ocean floor spreads away from the ridge. As a result, the ocean floor can be dated by its proximity to a ridge—the further from the ridge, the older it is. From this you might think that there must be patches of ocean floor that date all the way back to the origins of the Earth. But this cannot be so, because if it were, the Earth would slowly swell up, like a balloon, and yet (so far as we know) the Earth has remained approximately the same size throughout its history. As it turns out, the oldest known ocean floor dates back only as far as the Jurassic period. So what happened to the floors of all the oceans before that? These have been recycled by a mechanism that compensates for the production of rock at mid-ocean ridges, in which old ocean floor plunges into the depths of the Earth in ocean trenches. The result is a series of distinct areas of the Earth demarcated by ridges and trenches, from which rock is continually produced at ridges, and eliminated in trenches.

SHIFTING CONTINENTS

As new continents form, so they may disappear. Continental rocks are eroded and weathered, creating basins of sedimentary rocks as well as deposits on the ocean floor. Sometimes a hotspot, or even the germ of a new mid-ocean ridge, forms underneath a continent, splitting it in two. If a continent should be at the edge of a plate, moving toward another continent on an adjacent plate,

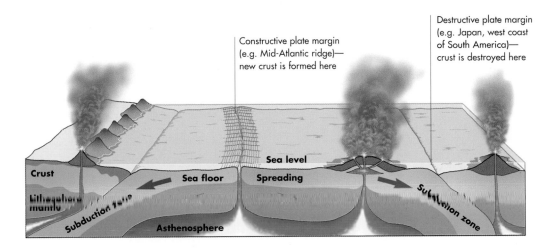

Constructive plate margin (e.g. Mid-Atlantic ridge)—new crust is formed here

Destructive plate margin (e.g. Japan, west coast of South America)—crust is destroyed here

Sea level

Crust

Lithosphere mantle

Sea floor   Spreading

Subduction zone

Asthenosphere

Subduction zone

LEFT New oceanic crust is continuously formed at mid-ocean ridges. When new crust is added to a plate that contains a continent, the continent is pushed away from the formative ridge. Over millions of years this activity has caused continents to move thousands of miles over the Earth's surface. At some plate boundaries, called subduction zones, oceanic crust is forced down into the Earth's interior, causing earthquake and volcanic activity.

N. Asia

LAURASIA

N. America   Europe

S. Asia

Pacific Ocean

Africa

Pacific Ocean

S. America   Tethys Ocean

GONDWANA

India

Antarctica   Australia

LEFT By the end of the Jurassic, around 145 million years ago, Western Laurasia has separated entirely from Gondwana and begins to resemble the familiar shape of the North American landmass. India, Australia, and Antarctica are also making their break. BELOW The continents in the late Cretaceous, 65 million years ago, are recognizably those we know today, with some adjustment to come. Australia is taking up its familiar location, and India is heading for a tumultuous rendezvous with south Asia.

N. America   Europe   Asia

N. Atlantic Ocean

Pacific Ocean

Africa

Pacific Ocean

S. America

India   E. Indian Ocean

S. Atlantic Ocean

Australia

Antarctica

the two bump into one another, creating huge mountain ranges.

Siberia, for example, is an ancient continental landmass formed billions of years ago from the accretion of many oceanic islands. In the Permian period, a collision between Siberia and the continent of Baltica threw up the Ural mountains. This process created basins in which abundant, fossil-rich sediment collected. Indeed, the Permian period gets its name from the city of Perm, in the southern Urals. At around the same time, this landmass joined with most of the others on Earth to create a single, giant continent, Pangea. This was slowly split by the creation of new mid-ocean ridges, but parts of them were to collide again, much later. Long after the extinction of the dinosaurs, the continental mass we call India collided with the southern edge of the Eurasian plate, creating the Himalayas. In this process, which is still going on today, the northern edge of India is slowly being pulled or "subducted" beneath Tibet.

Taking the long view, we see an Earth very different from the static conception that geologists once had. The continents are constantly on the move, oscillating between extremes of consolidation and fragmentation. Their movements have profound effects on life. The accretion of the continents to form one large landmass in the Permian reduced the area of the continental shelves available for marine life, and could have contributed to the phase of extinction at the end of the period. In the Jurassic period, for example, Antarctica was yet to adopt its place at the South Pole, and the North Pole was yet to be hemmed in by an almost-closed Arctic Ocean as it is today. The result was that warm water

was free to circulate even into polar regions, creating a uniformly warm, equable climate. As the continents assumed their more familiar positions, however, climatic variation became more marked. The high mountains produced by the collision of India with Asia disrupted global air circulation, creating the monsoon—and, perhaps, helping to nudge the Earth into a prolonged phase of cooling and drying that culminated in the Ice Ages in which modern humanity emerged. It seems likely that the fortunes of the dinosaurs were shaped by the movement of the continents, and in general by the slow break-up of the supercontinent of Pangea.

At the beginning of the Triassic period, the world was recovering from the extinction event at the end of the Permian in which more than 96 percent of all species of marine life was destroyed, along with a comparable number of species on land. It is possible that the center of Pangea, far from any ocean, became a vast, inhospitable desert. Once the world had regained its equilibrium, the Triassic saw the flowering of many species of land animal, including the earliest known frogs, turtles, and mammals, as well as many kinds of interesting, ultimately extinct reptiles. The dinosaurs were just one of these groups, and they first appeared toward the end of the Triassic. Most Triassic dinosaurs were small and bipedal, whether they were ornithischians or saurischians. At the end of the period, some kinds of dinosaur had become very large: The earliest sauropods appear at the end of the Triassic.

In the Jurassic period, dinosaurs in one place tended to look much like those in another, largely because the world's continents were still more or less bound together. The Cretaceous period led to greater diversity as the continents began to break up, splitting into something like the shapes we know today. Dinosaurs marooned on various landmasses went their own evolutionary ways, leading to distinctive, regional faunas. Hadrosaurs and ceratopsians, for example, are associated with eastern Asia and western North America, then linked together into a single island continent. Other kinds of dinosaur, such as the abelisaurid theropods, are associated with the southern continents. In general, theropods diversified to an astonishing degree, producing forms ranging in size from the gigantic Tyrannosaurus to the starling-sized Microraptor.

But perhaps the most important event to occur in the Cretaceous was the establishment of flowering plants as a feature of global ecology. Confined at first to lowland water margins (water lilies are among the most ancient kinds of flowering plant), they gradually spread across the landscape, diversifying as they went and creating entirely new kinds of forests, and, through those, new kinds of landscape. Along with the flowering plants came pollinating insects, and the germ of an essentially modern ecology.

THE END OF THE DINOSAURS

The apparent sudden demise of the dinosaurs is one of paleontology's great non-problems. Because it is impossible to forge a definite link between an extinction's possible cause and its consequences—especially for such a remote period in the Earth's past (after all, nobody we know was there to watch it happen), paleontologists have been free to come up with all kinds of reasons to explain why the dinosaurs died out.

For example, the weather became too hot, too cold, too wet, too dry, or a combination of the above. The dinosaurs were struck by some new disease; they succumbed to hay fever (from those new flowering plants), acid rain, volcanic eruptions, the radiation from supernovae, asteroid impacts, indigestion, or impotence. Perhaps the shells of their eggs became too thin, so they shattered prematurely—or too thick, so the hatchlings couldn't get out. Maybe the eggs and young fell prey to the depredations of newly evolved mammals. Or, perhaps, after such long residence as the kings of all beasts, the dinosaurs simply ran out of things to do and expired from boredom. All the above have been suggested, at one time or another, as the reason for the extinction of the dinosaurs. The candidate-du-jour is asteroid impact. It is certain that an extraterrestrial body about half the size of the island of Manhattan, traveling at tens of thousands of miles per hour, struck what is now the Caribbean coast of Mexico at around 65 million years ago, and that such an impact would have caused worldwide devastation. It is possible that the dinosaurs were among the most prominent casualties.

We have three problems with the idea of a mass extinction of the dinosaurs.

ABOVE It is certain that an asteroid hit the Earth around 65 million years ago causing immense devastation. However, evidence that the asteroid impact drove dinosaurs to extinction must remain circumstantial. After all, birds are specialized dinosaurs— and they survived.

The first is that it is never possible to link causes and effects in the way that many paleontologists assume to be possible. Until someone discovers a fossil of *T. rex* with a piece of asteroid between its teeth we will never know for certain that asteroids played any role in the death of any single dinosaur, let alone all of them, at once.

This leads us to our second objection—that extinction is conventionally seen as a unitary event, a scythe that cuts down an entire group at a stroke, when it is actually the summation of the deaths of many individuals, each one of which could have met its fate in a number of ways.

Third, extinction is a fact of life. Species become extinct every day. Mass extinctions represent an escalation of the background level of extinction, and the degree to which this escalation represents some new phenomenon, a "mass extinction," necessitating a search for some special cause over and above the slings and arrows of extinctions more generally, is a matter of interpretation.

However, we'd like to offer our own perspective on the extinction of the dinosaurs, and it has all to do with size. The end of the Ice Ages around 10,000 years ago saw the wholesale disappearance of most animals larger than a Golden Retriever. Until recently, large animals such as mammoths, bison, large species of deer, and so on were relatively common. Giant ground sloths roamed the Americas, and giant kangaroos hopped across Australia—but this is no longer true. It is possible, indeed likely, that emerging humanity was responsible for this slaughter, but irrespective of the cause, we suspect that large animals might be more prone to extinction than small ones. They are more visible and have fewer places to hide; they also tend to breed less frequently and have fewer offspring than smaller animals. By extinction, we think that whatever wiped out the dinosaurs preferentially affected the larger species, leaving the smaller ones—the birds—unscathed.

HOW TO USE THE FIELD GUIDE

The greater part of this book is arranged as a field guide, intended to be useful to safari enthusiasts with a fondness for time travel. We illustrate and describe a selection of dinosaurs as if they were living animals, with details of their appearance, habits, and most likely habitats, and notes on their ecology. We also provide handy thumbnail-style graphics for orientation in time and space. We have divided the Mesozoic into four phases: The Triassic, the Jurassic, the Early and Middle Cretaceous, and the Late Cretaceous. We have divided the Cretaceous for the sake of balance, given that probably as much is known about the dinosaurs of the last 20 million years of their existence as those of the previous 130 million. We detail the location of the dinosaurs by continent. For simplicity, these are given with reference to the modern continents. Although the world has changed markedly since the Mesozoic, it was beginning to be recognizable to our modern eyes, and the distributions of different kinds of dinosaur can, if roughly, be described in terms of modern landmasses.

Set in their natural habitat, Luis V. Rey brings to life the world of the dinosaurs with vivid colors and detailed textures

A sketch of the skeleton shows the detailed anatomy of each dinosaur

A global map pinpoints the region where each type of dinosaur can be found

A small cladogram, identifies the "family" that each dinosaur belongs to, as well as the period in which it lives

Anatomical sketches and notes detail the characteristics and habits of the dinosaurs

We'll finish with a warning to the reader: Those who choose to believe what follows do so at their own risk. Others will dismiss our speculations as preposterous and fantastical. And in one sense they will be absolutely right, for we are willing to bet that the real dinosaurs of the Mesozoic were not at all like anything you see in these pages—*they were far, far stranger.*

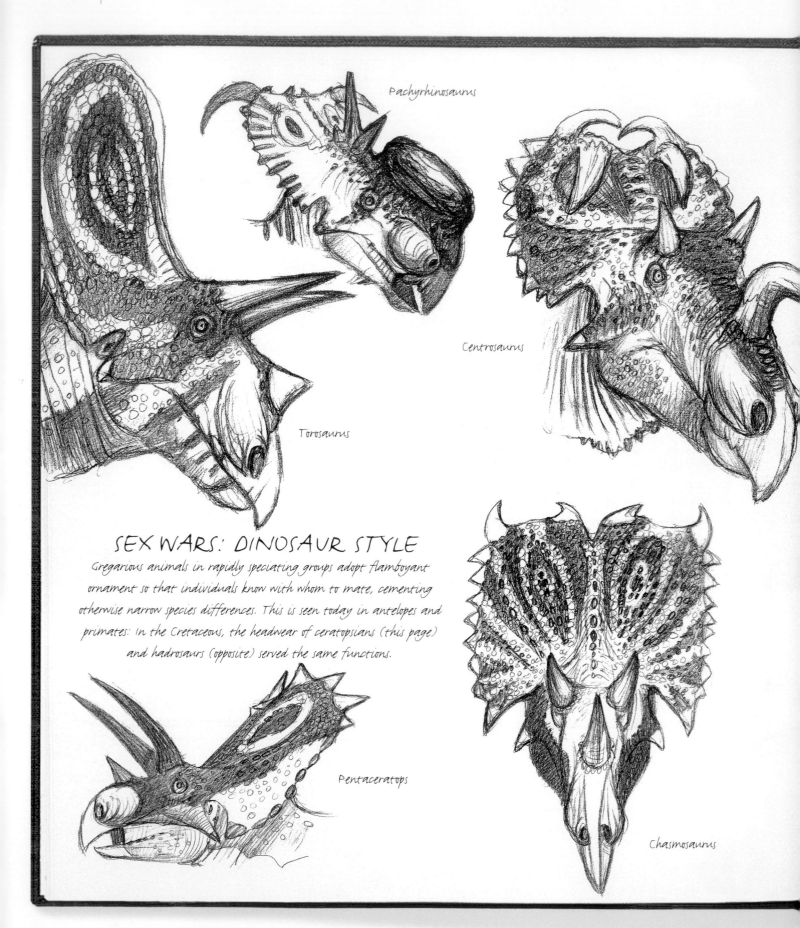

Pachyrhinosaurus

Centrosaurus

Torosaurus

SEX WARS: DINOSAUR STYLE

Gregarious animals in rapidly speciating groups adopt flamboyant ornament so that individuals know with whom to mate, cementing otherwise narrow species differences. This is seen today in antelopes and primates: in the Cretaceous, the headwear of ceratopsians (this page) and hadrosaurs (opposite) served the same functions.

Pentaceratops

Chasmosaurus

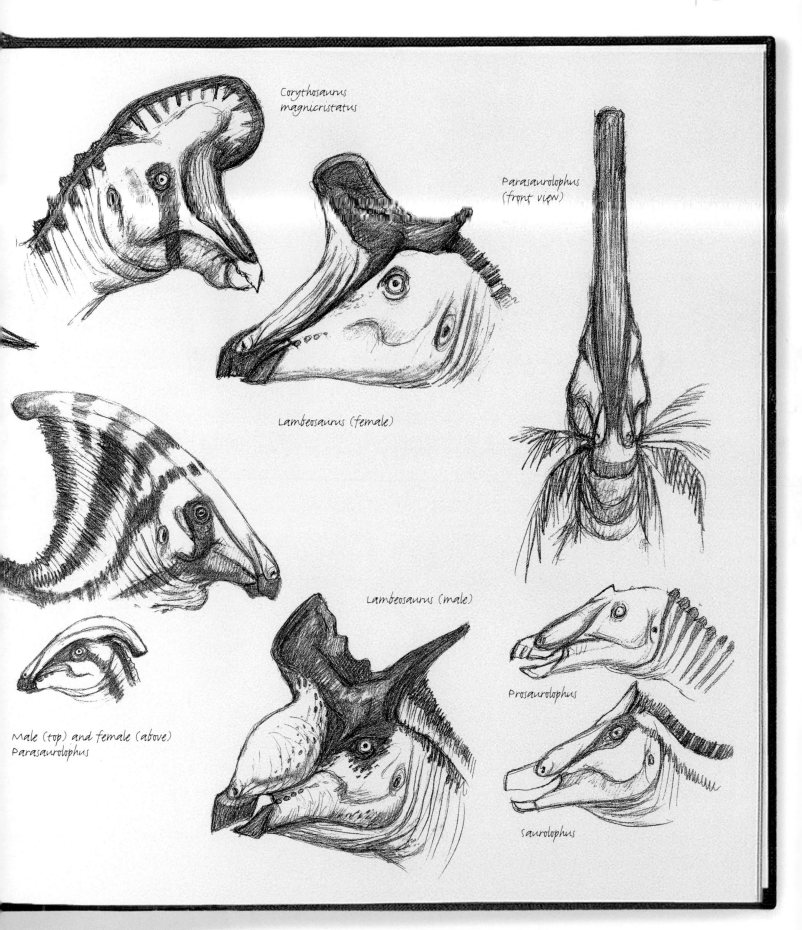

Corythosaurus
magnicristatus

Parasaurolophus
(front view)

Lambeosaurus (female)

Lambeosaurus (male)

Male (top) and female (above)
Parasaurolophus

Prosaurolophus

Saurolophus

The Triassic

245 to 208 million years ago

period

COELOPHYSIS

Description: Small primitive theropod
Length: 7–14 ft (2–4m) nose to tail

Distinguishing Features: Several species of Coelophysis are known: The one illustrated here is *Coelophysis bauri*. Swift and agile runners with extremely long tails, these dinosaurs are unusual for theropods of the period in that they are highly social. The animals are gray- to blue-green with a camouflage of emerald-green stripes, scarlet wattles around the face, and golden-yellow muzzles. Males and females are similar in size, though males tend to be slightly larger and more stockily built. Mating can take place at any time of year—a male "guards" a receptive female and tries to mate while warding off other suitors, but a female will usually mate with several males before laying a clutch of six to eight green eggs. The eggs are buried beneath a thin layer of vegetation and abandoned. The precocial hatchlings emerge after 4–5 weeks as miniature adults and immediately start foraging for small invertebrate prey.

Habit and Habitat: Coelophysis are invariably found in packs of between 40 and 80 individuals of both sexes and all ages. The fact that clutches of chicks may be fathered by more than one male has been advanced to explain the persistent sociality of the species. Unlike sauropod herds, however, Coelophysis packs have no real organization or dominance hierarchy, and individuals regularly fight among themselves over prey. An invasion of these creatures has been described by one traveler as "rather like an army of man-sized soldier ants on the move, spelling instant destruction for any small animals unlucky enough to break cover before the ravening tide."

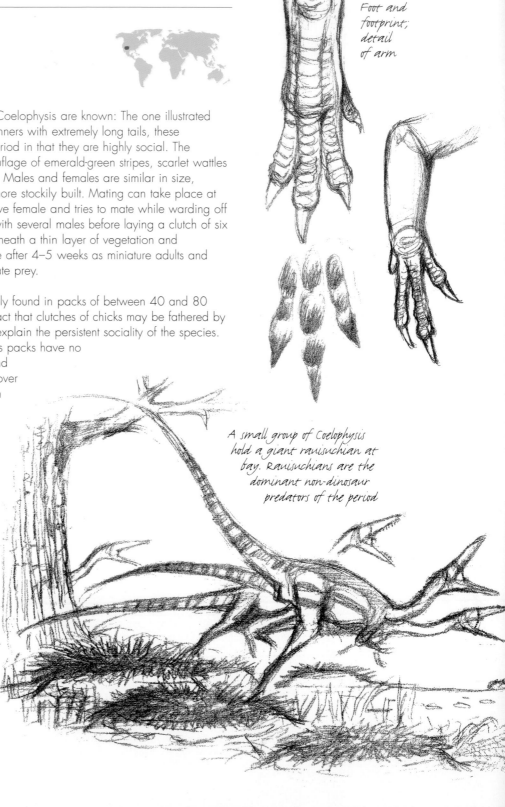

Foot and footprint; detail of arm

A small group of Coelophysis hold a giant rauisuchian at bay. Rauisuchians are the dominant non-dinosaur predators of the period

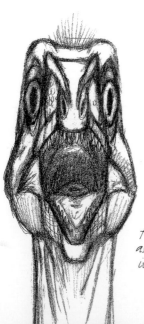

Coelophysis face-on, as seen by its prey!

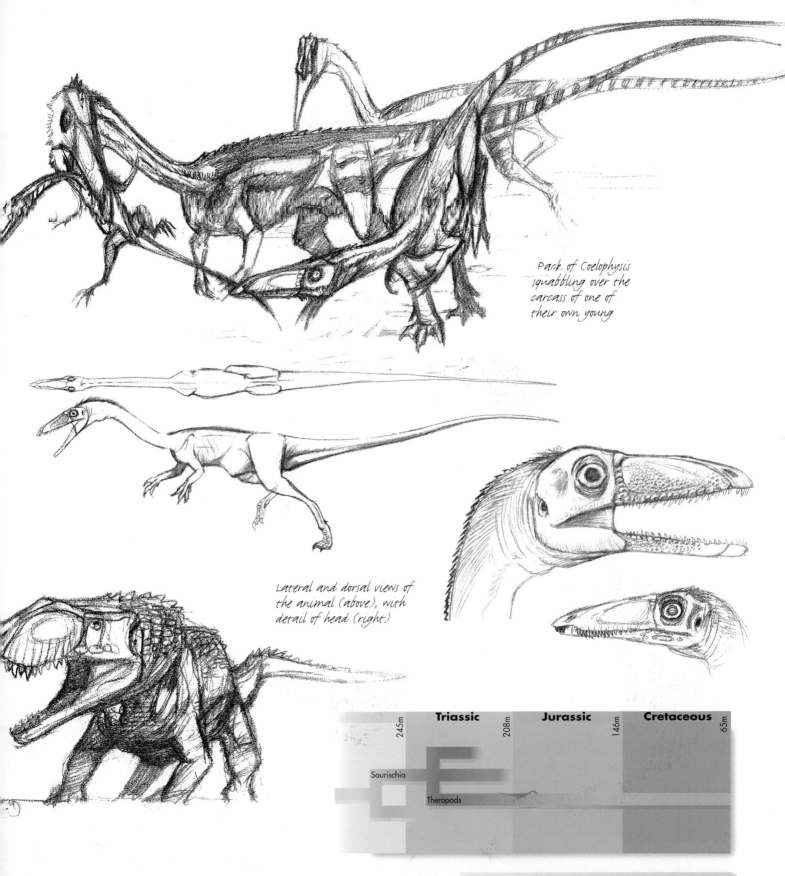

Pack of Coelophysis
squabbling over the
carcass of one of
their own young

Lateral and dorsal views of
the animal (above), with
detail of head (right)

		Triassic		Jurassic		Cretaceous	
	245m		208m		146m		65m
Saurischia							
Theropods							

Overleaf: Two adolescent Coelophysis males
fight over a freshly-caught lizard

EORAPTOR

Description: Small primitive dinosaur

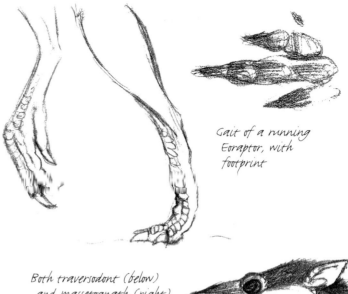

Gait of a running Eoraptor, with footprint

Distinguishing Features: This lightly built biped is reddish brown in color with lapis-blue markings on the head and neck and blue protofeathers on the flanks, particularly in males, during the breeding season. Males and females are similar in size, though females tend to be darker and duller in color and have less ornamentation. Courtship takes place in leks, in which one or more males display before a group of females, who then choose their mates. Females incubate clutches of 6–12 ellipsoidal, brown-flecked eggs in rough nests scraped in the ground.

Habit and Habitat: These animals roam open, poorly vegetated country from semiarid lowlands to heathland at higher elevations. Although Eoraptor will eat anything given the opportunity, small mammals form the major part of its diet. Eoraptor hunts alone or in small groups, invariably around dusk or just before dawn, when its small-mammal prey is most active. Paleontologists debate whether Eoraptor should be thought of as a generalized dinosaur or as a theropod in particular. Either way, it is one of the earliest known dinosaurs of any kind, already bearing the hallmarks of bipedal posture, grasping hands, large eyes, a fast-running metabolism, protofeathers, and relatively high intelligence. All these features suggest a specialism for hunting fast-moving, nocturnal prey, such as mammals. This may not be a coincidence, since the first recognizable mammals appeared at this time, opening an ecological niche for a new kind of predator.

Both traversodont (below) and massetognath (right) feature strongly on the Eoraptor menu

Three Eoraptor plunder the carcass of the dicynodont Dinodontosaurus

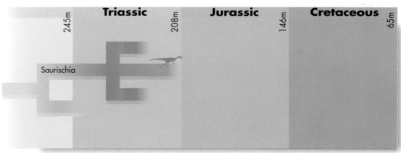

	Triassic		Jurassic		Cretaceous
245m		208m		146m	65m

Saurischia

Detail of hand to show vestigial 4th and 5th digits at far left

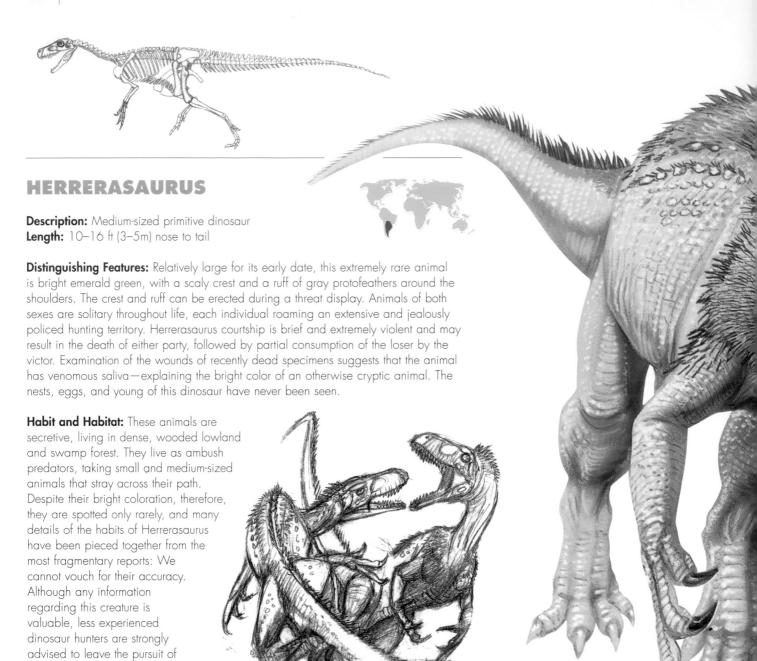

HERRERASAURUS

Description: Medium-sized primitive dinosaur
Length: 10–16 ft (3–5m) nose to tail

Distinguishing Features: Relatively large for its early date, this extremely rare animal is bright emerald green, with a scaly crest and a ruff of gray protofeathers around the shoulders. The crest and ruff can be erected during a threat display. Animals of both sexes are solitary throughout life, each individual roaming an extensive and jealously policed hunting territory. Herrerasaurus courtship is brief and extremely violent and may result in the death of either party, followed by partial consumption of the loser by the victor. Examination of the wounds of recently dead specimens suggests that the animal has venomous saliva—explaining the bright color of an otherwise cryptic animal. The nests, eggs, and young of this dinosaur have never been seen.

Habit and Habitat: These animals are secretive, living in dense, wooded lowland and swamp forest. They live as ambush predators, taking small and medium-sized animals that stray across their path. Despite their bright coloration, therefore, they are spotted only rarely, and many details of the habits of Herrerasaurus have been pieced together from the most fragmentary reports: We cannot vouch for their accuracy. Although any information regarding this creature is valuable, less experienced dinosaur hunters are strongly advised to leave the pursuit of this potentially lethal creature to the experts.

		Triassic		Jurassic		Cretaceous	
	245m		208m		146m		65m
Saurischia							

A violent courtship battle ensues as a male tries to mount a female (above). Not long after this scene the female killed and ate the male, leaving remains to be picked over by a group of scavenging Eoraptors

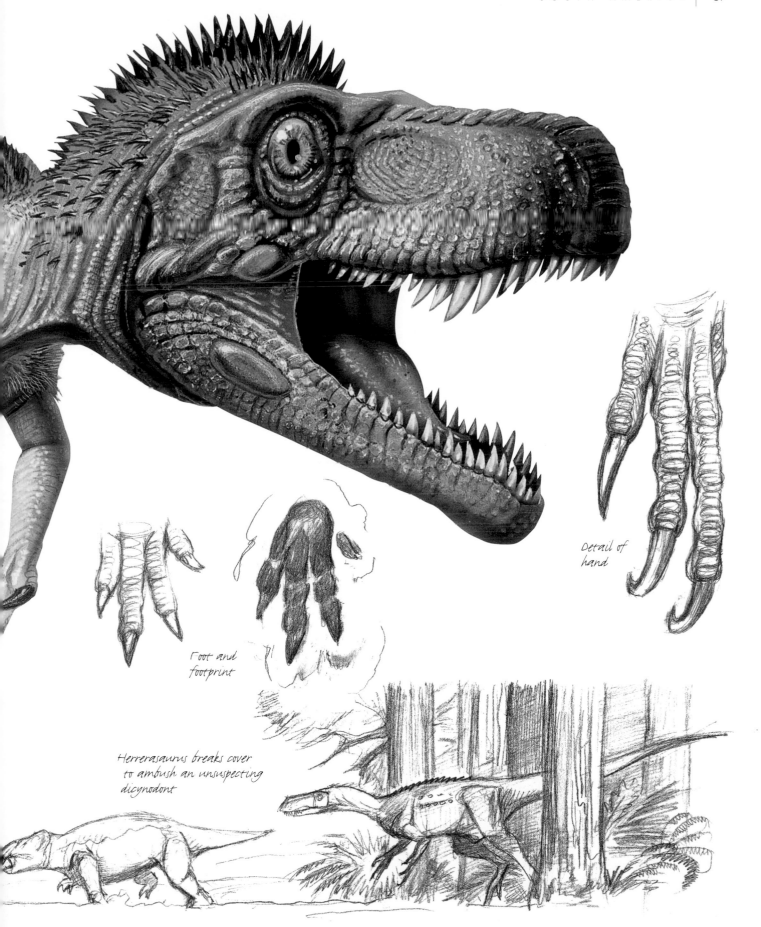

Detail of
hand

Foot and
footprint

Herrerasaurus breaks cover
to ambush an unsuspecting
dicynodont

LILIENSTERNUS

Description: Medium to large theropod
Length: 20–26 ft (6–8m) nose to tail

Distinguishing Features: This slender, gray-blue theropod has a prominent blue crest of scales down its back, flanked by blue and blue-black filoplumage which is generally longer in males than in females. The snout bears parallel black and yellow crests that are used in displays. Individuals usually live in small groups of either bachelor males or females with immature offspring. These groups coalesce into larger groups during the annual breeding season, when males develop extensive plumage on the body and arms and display noisily to the females. Unusually for dinosaurs, the females disperse before laying their eggs in small scrapes on the ground. Two or three precocial young hatch from a clutch of four or five eggs.

Habit and Habitat: A large theropod for the Triassic, Liliensternus hunts bigger prosauropods, early sauropods, and other herbivores in small groups consisting of either bachelor males or females and subadult offspring. The animals roam over a wide area, and each group monitors the movements of a number of herbivore herds. The theropods will usually try to corner a young victim, separating it from its herd and wearing it down by biting its neck and underparts.

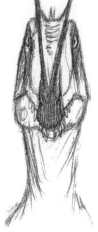

Liliensternus foot, showing reduced fourth digit. The fourth and fifth digits are almost or completely absent in later theropods to give the classic three-toed profile and footprint

Face-on profile of Liliensternus

Liliensternus (top) compared with Coelophysis, a smaller Triassic theropod

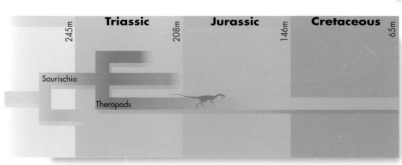

A pair of Liliensternus corner the small armored dinosaur Scutellosaurus

	Triassic	Jurassic	Cretaceous
245m	208m	146m	65m

Saurischia

Theropods

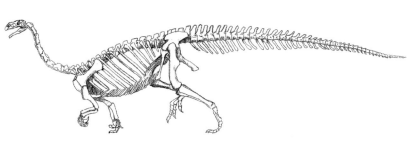

PLATEOSAURUS

Description: Medium-sized prosauropod
Length: 20–35 ft (6-10m) nose to tail

Distinguishing Features: There are many very similar species of Plateosaurus, distinguished largely by facial markings, coloring on the neck and flanks, and scale and plate ornamentation. The species illustrated is *Plateosaurus engelhardti*. Males and females tend to be similar in appearance, but females may be 10–20 percent larger than the males. The face is bright red, with red stripes continuing along the sides of the neck. The dorsal surface is tan to chocolate brown and rough in texture; the flanks, belly, and limbs are gray. Chicks are much more variegated in color, with zebra stripes of tan and chocolate.

Habit and Habitat: Like most prosauropods of this era, Plateosaurus lives for much of the time in family herds of between 5 and 20 animals grouped around a central matriarch and congregating into larger herds, sometimes of up to 200 animals, during the brief late spring breeding and mating season. The family groups separate and migrate to nesting grounds at higher elevations. Mating is polyandrous: Females in the established matriarchal hierarchy have access to a number of males proportional to their position in the pecking order. A mature dominant matriarch is about 30 years old and has three to five attendant male consorts, each of which guards a nest of 10–20 eggs, while the females perform general guard duties. The high-elevation, wooded nesting grounds are known as "fortresses." The dinosaurs often nest near trees containing roosting rhamphorynchoid pterosaur "sentries." These locations offer protection from flooding and easier defense against predatory theropods.

Male attends nest, turning the eggs between incubation bouts

Plateosaurus eggs are varied in size

Males compete for the attentions of the dominant female

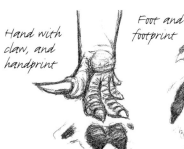

Hand with claw, and handprint

Foot and footprint

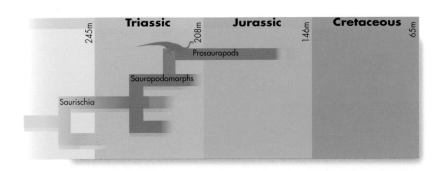

	Triassic		Jurassic		Cretaceous	
245m		208m		146m		65m

Prosauropods

Sauropodomorphs

Saurischia

Right: Roused by the screeching of the Pterosaur sentries, a Plateosaurus matriarch rises on her hind legs and displays her fearsome claws, ready to defend the fortress against a pair of marauding Liliensternus

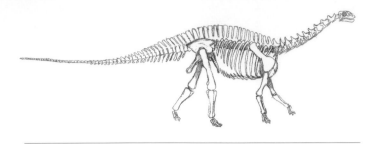

Isanosaurus head (right) showing teeth and typical sauropod leaf-cropping action, with Plateosaurus head (left) for comparison

ISANOSAURUS

Description: Small primitive sauropod
Length: 18–35 ft (5–10m) nose to tail

Distinguishing Features: Males and females are similar in appearance, though females are 10–20 percent larger than males, which are more vividly colored. The main illustration shows a male feeding from a conifer, with a female in the middle distance. Living largely in lowland coniferous woodland and flooded swamp forest, the animals show a cryptic coloration of green stripes. Isanosaurus is the earliest known true sauropod, although the differences between this form and the many contemporary species of prosauropod (see Plateosaurus) concern anatomical details rather than habit or appearance. Isanosaurus hands and feet have five, clawed toes, rather than the reduced number of claws seen in many prosauropods and sauropods. However, the sauropod-like calluses on its heels betray a certain heaviness and a tendency toward a more quadrupedal habit than is typical of prosauropods.

Mating pair of Isanosaurus

Habit and Habitat: Depending on the openness of the terrain, Isanosaurus is found either in temporary monogamous pairs or family herd groups dominated by a long-lived matriarchal hierarchy around which cluster a large number of fissile bachelor groups. Isanosaurus shows the tendency toward longevity, as well as the marked gender-related difference in age characteristic of sauropods. A dominant female can live for 30–50 years, while most males live for 20–25 years.

Hands and feet, with hand and footprints

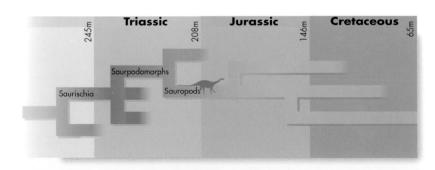

	Triassic		Jurassic		Cretaceous	
	245m	208m		146m		65m

Saurischia

Saurpodomorphs

Sauropods

The Jurassic

208 to 146 million years ago

period

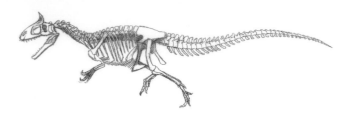

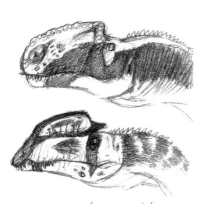

A female guards a nest on the upper shore while her mate forages offshore. The eggs are three-quarters buried in the sandheap close by

CRYOLOPHOSAURUS

Description: Medium-sized theropod
Length: 16–26 ft (5–8m) nose to tail

Distinguishing Features: A primitive member of the group of "advanced" theropods that includes tyrannosaurs as well as birds, this animal is found in the cool-temperate coastal regions of Antarctica and in other parts of southern Gondwana. Its chief distinguishing feature is a large and ornate head crest, found in both sexes but more brightly colored in males during the breeding season. The crest in *Cryolophosaurus elliotti*, the species shown here, is lemon yellow with blue bars, but colors vary among species. The trunk is clothed in a light pelt of black and white protofeathers, and the face may have a mask and beard of black pelage. The head is powerful, with deep, crushing jaws suitable for the seafood diet favored by this species. Breeding is highly seasonal and takes place in the early spring, coinciding with maximal nutrient upwelling in the Southern Ocean and abundant resources for feeding the young. Cryolophosaurus mates for life, usually nesting on the same beaches throughout life. The nests, made from heaped sand on the upper shore above the tideline, contain as many as twenty eggs, and are guarded by one parent while the other forages. The young are precocial and are sexually mature within two years.

The unusual head crest of Cryolophosaurus is not unique among theropods, as shown in these sketches of Monolophosaurus (top) and Dilophosaurus

Habit and Habitat: Cryolophosaurus is a specialist beachcomber of the ocean margins, picking up such *fruits de mer* as the carcasses of turtles, crocodiles, mesosaurians, and plesiosaurs. Its particular favorite is, however, ammonites, foraged from the strandline or the shallows. The shells of these giant molluscs are easily crushed by the dinosaur's deep jaws. The prey is swallowed half-crushed, and the meal stored in the gizzard for later regurgitation and consumption by the nestlings. Despite its fearsome appearance, Cryolophosaurus rarely preys on contemporary herbivores.

Two males guarding adjacent nests display to each other to emphasize territorial boundaries

The long, slender eggs are buried nose-down in sand. Their buff, mottled color provides effective camouflage

A male picks up a large ammonite from a shoal stranded at low water

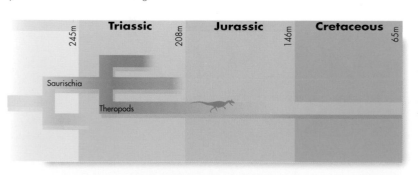

	Triassic	Jurassic	Cretaceous	
245m		208m	146m	65m

Saurischia

Theropods

MASSOSPONDYLUS

Description: Small to medium-sized prosauropod
Length: 10–16 ft (3–5m) nose to tail

Distinguishing Features: Like many prosauropods (see Plateosaurus), Massospondylus is brightly colored, although this allows for camouflage under certain circumstances. In the case of Massospondylus, the blue and yellow livery provides excellent cover for a life close to beaches and low-growing scrubland among sand dunes. Males and females differ in proportion: Females are slightly larger than males and have longer necks. As with Plateosaurus, chicks are highly variegated in color, allowing them to blend in with nesting-site scenery.

Habit and Habitat: Massospondylus, a late member of its group, bucks the trend of its sauropod cousins in having less social organization rather than more. Animals live in loosely organized matriarchal groups that congregate into larger herds for the spring mating season. Males compete for females, but there is no long term social structure. Females lay six to ten eggs in small nests about 1 yard (0.9m) square under sand and vegetation; the chicks are precocial and are ready to move with the herd within a few days of birth. These dinosaurs live on tough, shrubby plants but also scavenge for shellfish and carrion, and dig for invertebrates and roots with their stout front claws. Strength in numbers, as well as their front claws, offers protection against theropod predators.

A foot (top) and a hand of Massospondylus, with corresponding prints

Massospondylus on the move

Massospondylus taking a conifer cone

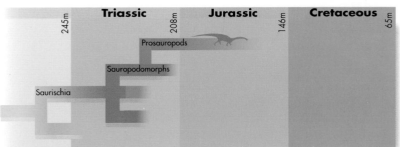

	Triassic		Jurassic		Cretaceous	
245m		208m		146m		65m

Prosauropods
Sauropodomorphs
Saurischia

Overleaf: Rearing up and brandishing their claws, male Massospondylus deter a foraging Cryolophosaurus

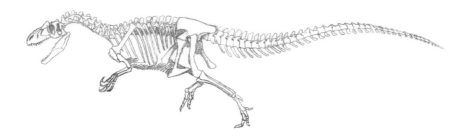

ALLOSAURUS

Description: Medium to large theropod
Length: 30–40 ft (9–12m) nose to tail

Distinguishing Features: This is a lightly built, fast-running, pack-hunting theropod. Color and ornamentation vary widely between seasons and from species to species. The most common species, *Allosaurus fragilis*, generally shows a mottled yellow-green "combat camouflage" pattern on the back, neck, and tail and is a dull gray on its limbs and underparts. The top of the head is usually ornamented with bony sculpture and arrays of small horns, the precise pattern of which can be used to identify individuals. Males and females look very similar and are notable for their general lack of plumage or ostentatious scale ornamentation—presumably an adaptation to streamlined pursuit. The exception is observed during the brief springtime breeding season, when males assume luxuriant plumage, including showy crests and wattles, and long, vaned feathers on the forearms and tail, displaying noisily before females in communal leks. The species is monogamous; males and females raise a brood of three to four chicks in well-protected nests on high ground.

Habit and Habitat: Groups of related *A. fragilis* tend to share hunting territory which they patrol in loosely cooperative packs of between four and ten individuals. They pursue and harry their prey for long distances, raking the hindquarters of the prey with slashing bites designed to hinder rather than to kill immediately. The hunters pounce only when the prey is close to exhaustion and can be dispatched easily. Their main targets are iguanodontids such as Camptosaurus, and stegosaurs. They compete for prey with other, smaller theropods such as Ceratosaurus, which they will occasionally drive from their kills, and they will also scavenge. *Allosaurus fragilis* is one of the smaller theropods capable of mounting attacks on large sauropods, though such targets are relatively rare. These Jurassic giants are, however, a specialty of the closely related *Allosaurus maximus*. This very large creature (up to 52 ft [16m]) is rarely seen by virtue of its exclusively nocturnal habit and its permanent coat of midnight-black pelage. *Allosaurus maximus* inflicts significant damage on encampments of migrating sauropods, though this can usually be assessed only from the carnage that greets the dawning day.

The large, creamy-white egg of A. fragilis, next to a week-old hatchling covered in downy feathers and camouflage colors

from the side, and
from the front to
show the gape

Allosaurus fragilis in left
lateral and dorsal view

Torso of running
A. fragilis seen from
the front. The palms
of all theropods face
inward as they move

	245m	**Triassic**	208m	**Jurassic**	146m	**Cretaceous**	65m
Saurischia				Allosaurs			
		Theropods					

DIPLODOCUS

Description: Large sauropod
Length: 70–100 ft (20–30m) nose to tail

Distinguishing Features: The first thing the dinosaur hunter notices is the extreme length of this dinosaur compared with its relatively slight build. More than two thirds of the almost 100-feet (30-m) length of this animal is taken up with neck and tail, held out more or less horizontally. Diplodocus is additionally marked by a head-to-tail crest of triangular plates. These grow throughout life—older individuals have longer plates. Males and females look very similar, though females tend to be larger. Coloration is generally gray, grading into reddish pink on the flanks, underside, limbs, neck, and face. The neck and tail are variously patterned with pink and gray stripes. Twenty to thirty eggs are laid in parallel grooves scraped in the earth and covered with vegetation and dung and guarded by a male or a low-rank female related to the layer.

Habit and Habitat: Typifying sauropod social organization at its height, Diplodocus lives in fluctuating but large (20–100 individuals) family-based clans, characterized by a matriarchal pecking order dominated by a long-lived queen with her close female relatives and a pack of consorts, each of which attends one of several nests laid by a single female. Longevity is as extreme as the physical length of the animal—Females can live for 100–120 years depending on social status, and males usually not much less than a century. Diplodocus herds tend to remain coherent throughout the year; they are largest when congregating in lowland forest and plains, though smaller bands occasionally forage in woodland along the upper slopes of river valleys.

Diplodocus eggs

Only a juvenile Allosaurus would be foolish enough to approach an adult female Diplodocus tending her young. One lash of her whiplike tail can deliver a fatal blow

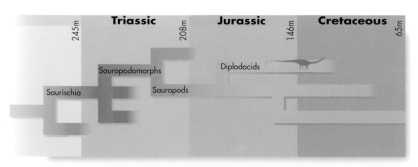

Triassic — 245m — 208m — Jurassic — 146m — Cretaceous — 65m

Saurischia
Sauropodomorphs
Sauropods
Diplodocids

Two males fight to gain female attention. Using their tails and both hind legs as a tripod, they rear up before each other—a stance that cannot be maintained for long for risk of fainting. The combatants try to gash the tender skin of their opponent's throat with the spiny plates on the back of their neck

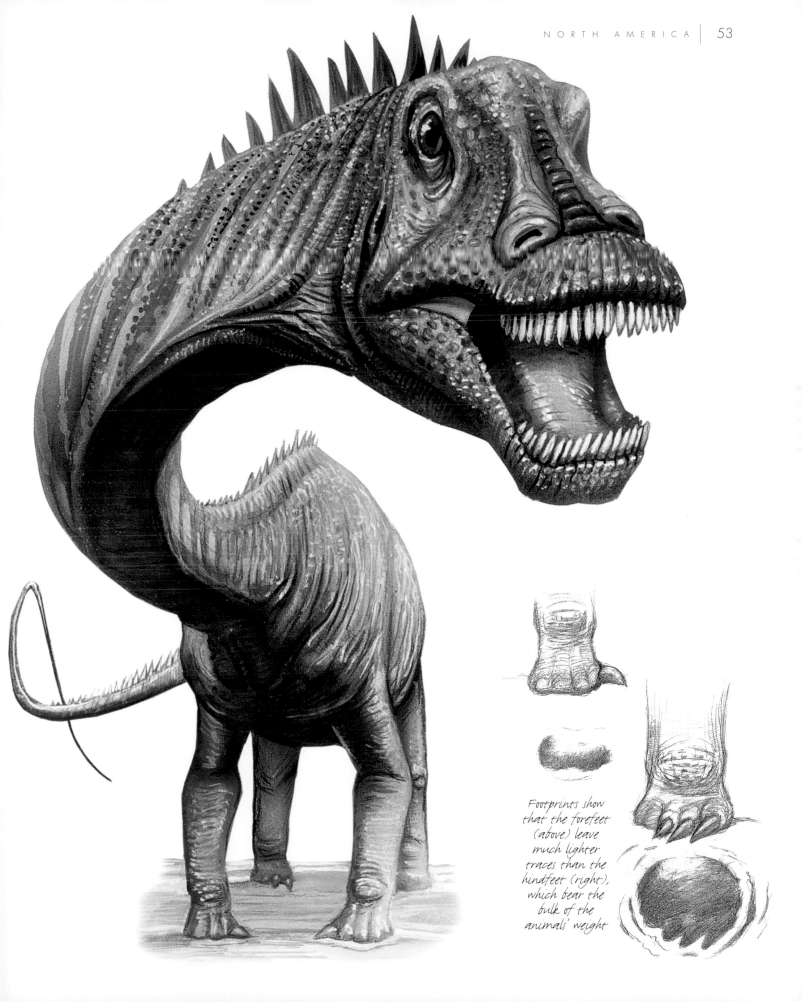

Footprints show
that the forefeet
(above) leave
much lighter
traces than the
hindfeet (right),
which bear the
bulk of the
animals' weight

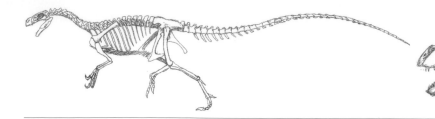

ORNITHOLESTES

Description: Small theropod
Length: 6.5 ft (2m) nose to tail

A female Ornitholestes awaits her mate in receptive posture

Distinguishing Features: This is a small, often brightly colored theropod and one of the few that comes close to being a specialist egg robber. Males (as shown in the main illustration) have fluffy, white plumage and golden-yellow snouts which are accentuated during the spring-summer mating season with a bright red crest and blue, red, and black spots and stripes. Females have off-white plumage with occasional black or gray markings, but the head is bald and brick red, reminiscent of that of a vulture. Males and females pair for life and raise a brood of three or four young each year. Pairs of Ornitholestes are often found on the fringes of sauropod nesting grounds and adjust their breeding season to accommodate that of their "host." Usually, each species of sauropod is "parasitized" by a single species of Ornitholestes. Out of season, and after the young are fledged, these animals tend to forage alone, tracking sauropod herds and feeding on whatever they can find.

Habit and Habitat: Widely distributed but not common locally, these animals are exclusive associates of sauropod herds, as mentioned above, and have become an important vector for the elaborate life cycle of a remarkable parasite, the fluke *Praefasciola brachiosauri*. The adult worms—each of which may reach 10 feet (3m) in length—live in the immense livers of female sauropods where they mate and lay millions of eggs. These eggs transform immediately into microscopic larvae called procercariae and these make their way to the sauropod oviducts, where they infect eggs as they form. Each procercaria transforms into a bag containing dozens of cercariae, the next larval stage which form dormant cocoons. As such they would die if the eggs were not eaten by the right species of Ornitholestes. Once inside the predator, each of the cercariae breaks up into thousands of single-celled metacercariae, which find their way into the animal's bloodstream. The final stage involves a fly, *Luisreya ginsbergi*, which lives on blood meals collected from the nasal passages of dinosaurs. The cycle is completed with the transmission of a parasite-loaded blood meal from an infected Ornitholestes to a female sauropod, in which the single-celled metacercariae develop into gigantic adult worms. It has been suggested that the life cycle of the parasite not only exploits but also reinforces the association between Ornitholestes and sauropods, and that it may indeed be responsible for this relationship, and for promoting the evolution of a specialist egg robber from an otherwise rather generalist theropod.

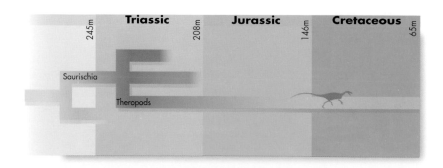

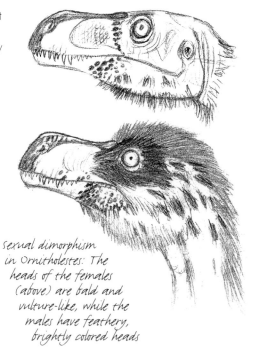

sexual dimorphism in Ornitholestes: The heads of the females (above) are bald and vulture-like, while the males have feathery, brightly colored heads

Ornitholestes making off with a Brachiosaurus egg

The animal has a dramatic display bluff, in which it stands almost vertically on tiptoes, stiffening the whole body and fluffing the feathers to appear bigger than it really is, while opening the arms and showing claws

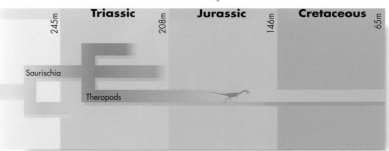

The flat nose-horn of Ceratosaurus is only incipient in hatchlings but is still prominent enough to help them break out of the egg

CERATOSAURUS

Description: Small to medium theropod
Length: 13–23 ft (4–7m) nose to tail

Distinguishing Features: Ceratosaurus is a pack-hunting theropod distinguished by elaborate head ornamentation. Many species have large nasal horns, as well as horns immediately in front of, or surmounting, the orbits. These horns are invariably brightly colored. *Ceratosaurus nasicornis* (pictured) has bright-red horns, as well as a generally bright-red neck, back, and tail, with horizontal red stripes on the flanks over a dull gray background. The short, powerful neck and back are relatively well armored with bony nodules and scutes. Plumage is not present, except in juveniles and—during the breeding season—in males, where it develops ostentatiously for competitive display in leks. Breeding habits are similar to those of Allosaurus and, indeed, many medium-sized theropods of the period. The teeth are long, even for theropods, and the gape is very wide, enabling a saber-tooth slashing action. Males and females tend to pair for life, re-establishing nest sites each year. The chicks are precocial, but parents care for their young for several months after birth, teaching them to hunt.

Detail of claws

Like most theropods, the chicks of Ceratosaurus are covered in downy plumage and have a cryptic pattern of stripes or mottling. Here, the precocial chicks learn hunting with their mother

Habit and Habitat: Ceratosaurus hunt in packs of three or four animals, usually either exclusively males or females. They tend not to be pursuit predators, however, preferring to ambush their prey. Iguanodontids are relatively easy pickings, though *C. nasicornis* is a specialist in armored dinosaurs such as the *Stegosaurus armatus* shown on the next page. Although slow and unintelligent, stegosaurs are ferocious when cornered, which perhaps explains the protective horns and armor in their chief persecutors. As with Allosaurus, several species of Ceratosaurus exist, and some of the larger ones specialize in sauropods. *Ceratosaurus ingens* from Africa is much larger than *C. nasicornis* and subsists exclusively on very large sauropods such as Brachiosaurus.

The powerful teeth, short neck, and wide gape of Ceratosaurus make it capable of inflicting terrible shark-like wounds on its fleeing prey

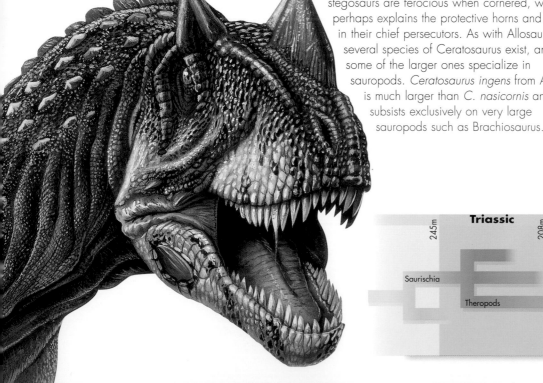

	Triassic	Jurassic	Cretaceous
245m	208m	146m	65m

Saurischia

Theropods

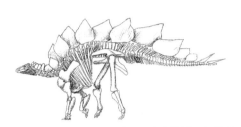

STEGOSAURUS

Description: Large armored dinosaur
Length: 26-30 ft (8-10m) nose to tail

Distinguishing Features: The largest of the "plated" dinosaurs, *Stegosaurus armatus* (illustrated here) is a bulky quadruped, generally a variegated green color but made unmistakable by the line of broad bony plates along its back. These plates, arranged in a staggered, double row, may be colored and patterned in various ways believed to be connected with species and individual recognition, and help to break up the outline of the slow-moving animal when seen against vegetation. They also serve to deter predatory theropods from leaping onto the flanks of the animal. The tail terminates in a bunch of four robust spikes, which can be waved from side to side. Stegosaurs live in small family groups dominated by a single matriarchal female, although social organization is much less elaborate than in sauropods. Groups come together annually at the rutting and breeding grounds, when animals—especially unmated males— move between family groups. Rutting males have brightly colored throat patches. The animals breed cooperatively, the larger males and females forming sentries to guard dozens of nests, each containing 10–12 eggs.

Habit and Habitat: This peaceable herbivore browses on low, soft vegetation at forest margins and along river banks—anywhere that there is likely to be an abundance of young, fresh vegetation. The trampling action of herds of stegosaurs incidentally creates the conditions in which young plants flourish: Stegosaurs tend to move from one feeding ground to another in a cycle that takes between 3 and 6 months, so that vegetation is always in its prime when the herd reaches any given site. A stegosaur also takes invertebrates, small mammals, eggs, and some carrion, all of which it swallows whole and grinds up in a gizzard—for which purpose it also swallows river gravel and pebbles.

stegosaurus loses a tail spike in a Ceratosaurus attack

stegosaurus tail spikes seen from above

A male stegosaurus in rut, waving its tail, rearing up on its hind legs and displaying the underside of the neck, which may be brightly colored during the mating season. The lines of the limb bones have been superimposed on this sketch

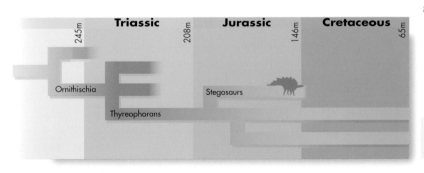

	Triassic		Jurassic		Cretaceous	
245m		208m		146m		65m

Ornithischia

Thyreophorans

Stegosaurs

The three-toed hind foot and four-toed forefoot. Only two of the toes on the forefoot bear claws

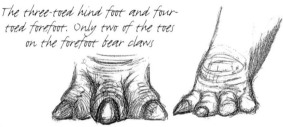

Overleaf: *Three Ceratosaurus slowly encircle a stegosaurus, which raises its tail in defense*

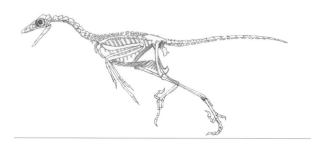

A female uses flapping flight to assist vertical escape up a tree trunk

Details of claws on wings, and feet adapted for running

ARCHAEOPTERYX

Description: Small flying theropod
Length: 12–24 in (30–60cm) snout to vent

Distinguishing Features: The several known species of Archaeopteryx, a feathered theropod dinosaur, are variable in color, habits, and habitat. This description is for *Archaeopteryx lithographica* (pictured). The feathers are deep blue to slate gray. The snout is gray; the wattles around the face are malachite green; and the underparts, legs, and feet are red. Females are on average slightly larger than males, but males have more prominent head plumage. The animals continue to grow throughout life, so both sexes vary greatly in size with age. They roost and forage in montane forest during the winter dry season, returning to open woodland around lakes and river margins in late spring to breed. Pairs are usually monogamous, though some extra-pair copulation has been observed in the crowded breeding territories. Nesting haphazardly in low shrubs or on the ground, a pair incubates four to six bright-blue eggs. Hatchlings are precocial and largely white with black or chocolate-brown mottlings. They fledge at three months and can breed in two years.

Habit and Habitat: The subtropical to tropical range of Archaeopteryx coincides with that of a number of small theropods such as Compsognathus, many with feathers, although Archaeopteryx is one of the few that are competent fliers. Flight is used sparingly—most energetically for courtship and also for territorial fights in the breeding season. Most characteristic, however, is the looping, intermittent flight low over water at dusk, as flocks of Archaeopteryx chase swarms of small insects such as gnats and midges.

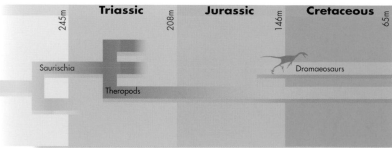

Heads of female
(left) and male
A. lithographica

Female incubates eggs
in sprawling nest in
low shrubs, while the
male looks on

Male leaps and
flaps over water in
pursuit of insects

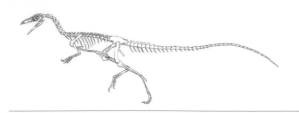

COMPSOGNATHUS

Description: Small primitive theropod
Length: 3 ft (1m) snout to vent

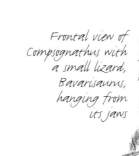

Distinguishing Features: This is a small, long-snouted theropod with a very long tail. Males and females are the same size but are quite different in appearance. Males have prominent black plumes on the head and midline of the back, and the rest of the body is patterned with peacocklike "eyes" with black spots fringed with white on a silver-gray background. Females lack plumes and are a uniform drab or brown. The animals are invariably found in large flocks. During the spring mating season, males build bowers of leaves, cones, and small, shiny objects (beetles, fish scales, and so on) as stages where they display noisily to an audience of females. The prospective mate moves into the bower, dismantles it and rearranges it into a nest. Clutches of six to eight eggs are incubated by both sexes, and the young are precocial.

Compsognathus is a cooperative breeder; that is, it is not uncommon to observe males displaying alongside their sons and other close male kin or to find several related, juvenile, or subadult animals helping to incubate a clutch.

Frontal view of Compsognathus with a small lizard, Bavarisaurus, hanging from its jaws

Males and females compared, showing the head-crest and piebald coloration of the male (top), contrasted with the uniform brown plumage of the female

Habit and Habitat: Found in locations ranging from lowland swamp forests to scrub, Compsognathus is never far from water, where it forages for invertebrates and small fishes. On land it takes small mammals, lizards, and the eggs and chicks of ground-nesting birds such as Archaeopteryx. Travelers to Jurassic Europe should always expect to acquire a retinue of these intelligent, inquisitive creatures—and should beware of losing small, shiny objects. Recent descriptions of Compsognathus bowers report such exotica as wristwatches, candy-bar wrappers, loose change, digital cameras and, in one case, an unexploded stun grenade.

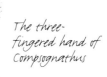

The three-fingered hand of Compsognathus

	Triassic		Jurassic		Cretaceous	
245m		208m		146m		65m

Saurischia

Theropods

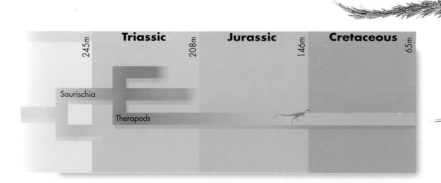

Compsognathus running down a small mammal

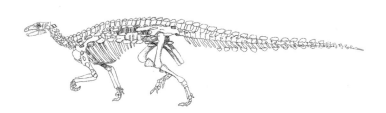

SCELIDOSAURUS

Description: Early armored dinosaur
Length: 8–15 ft (2.5–4.5m) nose to tail

Distinguishing Features: This relatively lightly built armored dinosaur is uniformly blue-gray in color. The dorsal surface is protected by seven parallel rows of pale-gray scutes, and additional scutes protect the forearms, neck, tail, and back of the head. The skin between the scutes, especially on the back, is tough and fibrous. Scelidosaurus is either solitary or found in long-lived male-female pairs. Unusually for dinosaurs, they have no particular breeding season and do not reproduce at all in most years. Occasionally they produce two clutches per year, each one with four or five eggs of which no more than two hatch. The young accompany the adults for 4 or 5 years before they venture out on their own. They do not attain sexual maturity for another 10 years. This careful strategy suggests that the animals may live to an advanced age. Indeed, some are believed to be more than 200 years old. The scutes in older animals are occasionally shed and replaced, but extra ones are added. The oldest animals have additional rows of scutes on their back, neck, and head, and a tail almost completely covered with scutes—rather like that of an ankylosaur in appearance.

Habit and Habitat: This dinosaur is typically found in extremely dense, lowland swamp forests and in mangroves, rivers, and estuaries where it browses on weeds, water plants, worms, and snails. It is a capable (if slow) swimmer, rather like a sedate vegetarian alligator. It can remain completely submerged for several minutes, and some reports claim that it walks along river bottoms. This shy, secretive animal has few predators apart from large crocodiles and the occasional pliosaur roaming upriver from the open sea.

Hand, showing how only two digits contact the ground. The claw of the first digit is kept clear, and the smaller fourth and fifth digits lack claws. The foot has four toes, all with claws

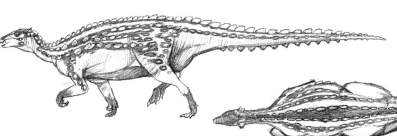

Left lateral and dorsal view of scelidosaurus, with (below) detail of scutes from the front, from the side, and from above

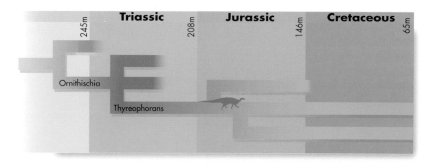

	Triassic	Jurassic	Cretaceous
	245m	208m	146m	65m

Ornithischia

Thyreophorans

BRACHIOSAURUS

Description: Large sauropod
Length: 65–105 ft (20–32m) nose to tail

Distinguishing Features: One of the largest dinosaurs—indeed, one of the largest animals that has ever existed—this sauropod is much bulkier than contemporaries such as Diplodocus and Mamenchisaurus. The forelegs are longer than the hindlegs, giving the back a high, sloping appearance, and the neck is habitually held more vertically than horizontally. The animal is gray to gray-brown, with brown mottling on the head, neck, dorsal midline, and shoulders. Red wattles on the forehead can be inflated during vocalization. Like many sauropods, this animal is highly gregarious, but among brachiosaurids, in contrast to other sauropods, society is male-dominated. A huge male dominates a harem of ten or more females which tend to be significantly smaller in size. A herd consists of the male, his harem and offspring, and a loose association of subordinate related males. The dominance of the alpha male is challenged in the spring mating season when subordinate males try to topple him with threat displays which can be extremely violent. The dinosaur watcher is advised to keep at a safe distance to avoid being deafened, squashed, or both. These contests are followed by mating, which is almost as spectacular (and as dangerous) as the fighting. Females lay a loosely arranged clutch of 10–12 eggs, which they cover with vegetation and guard.

Habit and Habitat: Brachiosaurus herds migrate between feeding areas in open upland coniferous forests, and breeding grounds in open parkland at lower elevations. These animals feed continually on tender, young leaves and cones, but this diet is, nutritionally, relatively poor, and consequently they must eat phenomenal amounts each day. Diet is supplemented by vitamins synthesized by symbiotic gut bacteria, which also help to digest tough vegetable matter. The effect of a large herd of giant sauropods on the environment is devastating. The herd must be on the move constantly so its members can keep feeding and so the woodland has a chance to regenerate. Brachiosaurus has few enemies apart from small theropods such as Ornitholestes that steal eggs and large nocturnal theropods such as *Allosaurus maximus*. A theropod attack, however, is rare, and the principal threat to an adult male Brachiosaurus is another male Brachiosaurus.

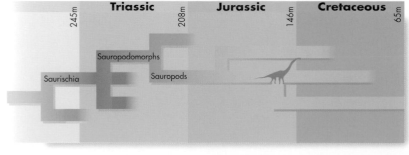

245m — Triassic — 208m — Jurassic — 146m — Cretaceous — 65m

Saurischia

Sauropodomorphs

Sauropods

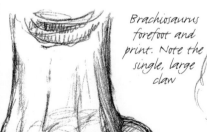

Brachiosaurus forefoot and print. Note the single, large claw

The leaf-shaped tooth of a Brachiosaurus

Lateral view of Brachiosaurus head

immature
Tuojiangosaurus
with relatively
small spikes and
plates

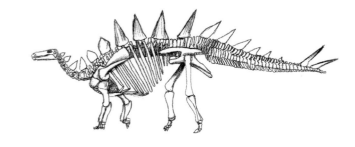

Head of
Tuojiangosaurus
(bottom) compared
with two other
armored dinosaurs,
Hesperosaurus
(top) and
Huayangosaurus
(middle)

TUOJIANGOSAURUS

Description: Medium to large armored dinosaur
Length: 20–26 ft (6–8m) nose to tail

Distinguishing Features: Smaller and darker in color than its contemporary Stegosaurus, Tuojiangosaurus bears 15 pairs of triangular plates in two files along its back. These plates are narrower, taller, and more spikelike than those of Stegosaurus. The rows of plates are arranged symmetrically; unlike the arrangement in Stegosaurus, the plates do not alternate. The tail bears two pairs of very long spikes, and another huge spike is mounted on each shoulder. Another difference between this dinosaur and Stegosaurus concerns patterning. Whereas the colorful plates of Stegosaurus may play a part in camouflage, as well as species recognition, the plates in Tuojiangosaurus have more of a defensive function, being sharply pointed and a uniform deep gray. This dinosaur is generally dull in color, patterned in alternating deep gray and purplish-brown stripes. These animals live alone or in small groups with no real social structure. They mate opportunistically: A clutch of six to eight eggs is laid in a scrape in the ground, buried under rotting vegetation, and abandoned.

Habit and Habitat: This shy, largely nocturnal creature is semiaquatic, cropping low-growing plants and weeds along overgrown water margins and in swamp forests. Like all stegosaurs, it is an opportunistic, omnivorous eater: It grubs for worms, catches small fishes and crustaceans, and scavenges for carrion. Its densely wooded and watery habitat offers protection against large theropods such as Yangchuanosaurus—a danger to which it is exposed when it comes onto higher and drier land to mate, lay eggs, or migrate to a new feeding ground. However, few animals can penetrate its fortresslike thicket of plates and spines, especially when it assumes a defensive posture with plates and spines pointed toward the attacker.

Tuojiangosaurus at bay. Under attack from a pair of Yangchuanosaurus, the stegosaur crouches porcupinelike on its front legs, defending itself with its shoulder spikes while waving its vicious tail spikes threateningly from side to side

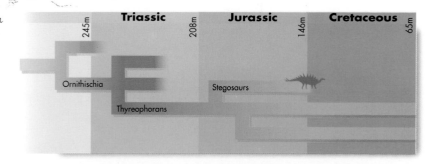

	Triassic		Jurassic		Cretaceous	
245m		208m		146m		65m

Ornithischia

Stegosaurs

Thyreophorans

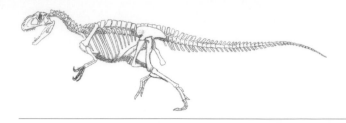

YANGCHUANOSAURUS

Description: Large theropod
Length: 30–40 ft (9–12m) nose to tail

Distinguishing Features: Larger than its contemporary Ceratosaurus and more heavily built than its relative *Allosaurus fragilis*, Yangchuanosaurus is more brightly colored than either, with alternating horizontal stripes of golden yellow and bright emerald green. In males the head is ornamented with elaborate bony scutes, generally golden in color. The slightly smaller females are duller and have less prominent head ornamentation. Like Allosaurus, this creature is monogamous, and a pair raises two or three fledglings at each nesting. Chicks are covered with a dense, downy, neutral-gray plumage. Such domesticity follows perhaps the most spectacular mating displays of any theropod. Just before the breeding season, males sprout a rich, peacock-blue pelage, an impressive feathery ruff and a fantastic array of iridescent peacocklike plumes on the arms, thighs, and tail. Thus arrayed, they preen and strut before an audience of females. Males do not feed during this period, and perform most of the parental care while females hunt in small groups of two or three. Mating plumage is quickly shed and provides ideal nesting material.

Habit and Habitat: Small groups of Yangchuanosaurus tend to live in lightly wooded lowlands, making camps in isolated spinneys on banks or hillocks that serve as combined nest sites, lookout posts, and home bases from which females hunt other dinosaurs. Being heavier than *A. fragilis*, these animals are built for steady tracking and ambush rather than pursuit. They specialize in large sauropods such as Mamenchisaurus, although they occasionally attack the stegosaur Tuojiangosaurus. The latter requires a cooperative strategy in which a pair of theropods distract the stegosaur from the front. While the prey mounts a threat display to its attackers, a third theropod attacks from behind, biting the unprotected cloacal region at the base of the tail.

A band of Yangchuanosaurus attacks a Mamenchisaurus that they have driven into a swamp

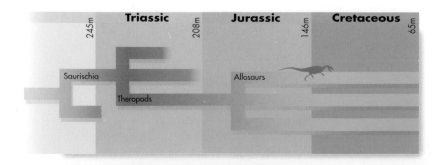

	Triassic		Jurassic		Cretaceous	
245m		208m		146m		65m
Saurischia			Allosaurs			
Theropods						

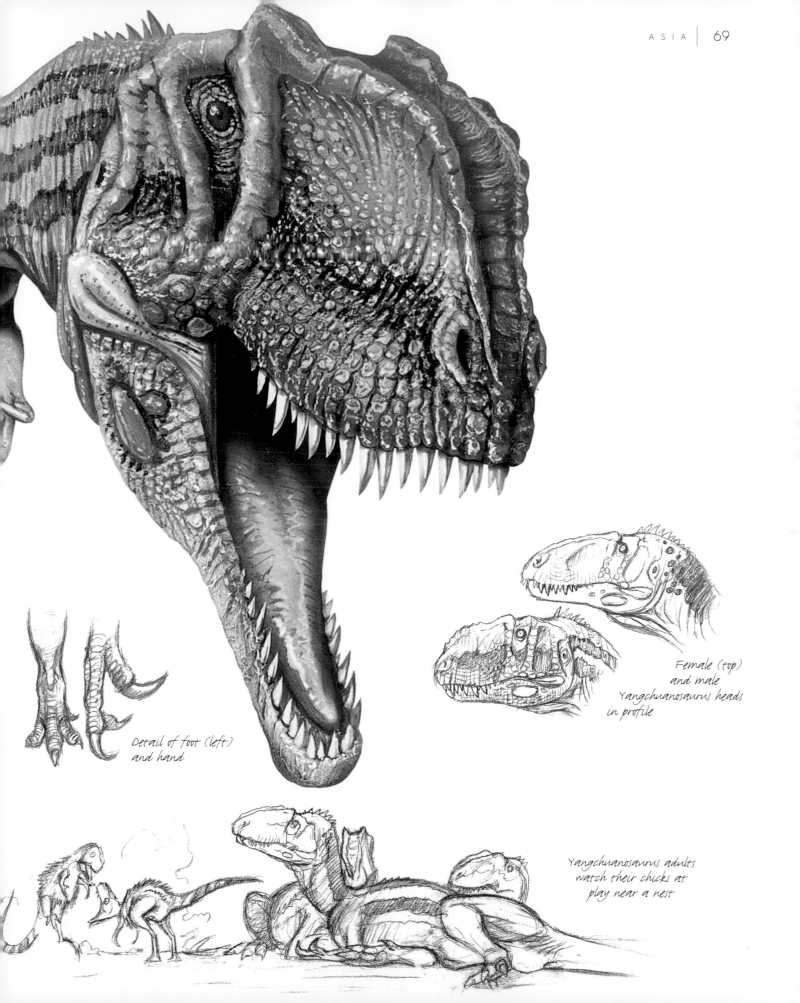

Detail of foot (left)
and hand

Female (top)
and male
Yangchuanosaurus heads
in profile

Yangchuanosaurus adults
watch their chicks at
play near a nest

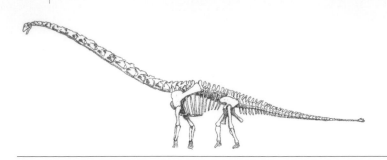

Detail of tail club

MAMENCHISAURUS

Description: Long-necked sauropod
Length: 65–85 ft (20–26m) nose to tail

Distinguishing Features: This animal is similar to Diplodocus in its relatively slight build and extreme length. The neck seems far too long and disproportionately thick for the slightness of the body. The tail terminates with a small bony club, and there may be a crest of raised plates along the midline. The limbs and flanks are mottled gray and the midline and upper surfaces of the body, neck, and tail are, in contrast, deep charcoal gray to black. These two regions are separated by a line of bright scarlet that extends from the jawline to the midtail. The nasal region and eyes may also bear red wattles, and the nasal wattle can be inflated during display or vocalization. A highly social animal, Mamenchisaurus is usually found in large herds of up to 100 individuals, formed around one or two matriarchal clans. Bachelor males are found in smaller groups congregating with the larger herds during the mating season. About 10–20 eggs are laid in a spiral fashion and then buried beneath vegetation and guarded until after hatching. The young grow very quickly and by early summer are big enough to move with the herd.

Habit and Habitat: These dinosaurs are mostly found close to water in valley bottoms and broad floodplains where their length allows them to graze far out over marshes and into open water. Their size and sociality mean they have few enemies. Only a few theropods such as Yangchuanosaurus are large enough to attack them, and if they do, the sauropods form a tails-out defensive ring around the herd, flicking their armored tails to great effect. Mamenchisaurus are also good swimmers, allowing effective escape from predators. This ability may also explain the occasional presence of large sauropods on offshore islands.

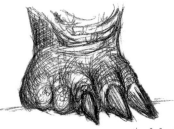

The large, spherical egg of Mamenchisaurus

Detail of foot

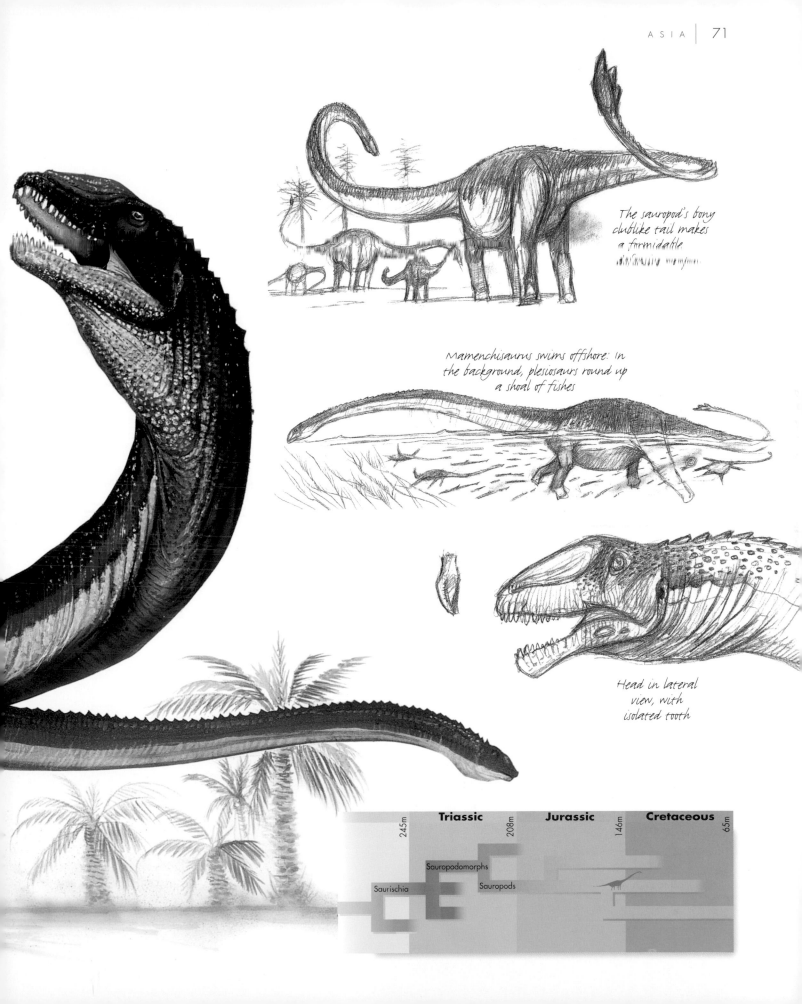

The sauropod's bony clublike tail makes a formidable.

Mamenchisaurus swims offshore: in the background, plesiosaurs round up a shoal of fishes

Head in lateral view, with isolated tooth

| 245m | Triassic | 208m | Jurassic | 146m | Cretaceous | 65m |

Sauropodomorphs

Saurischia

Sauropods

The earl
Cret

146 to 100 million years ago

und mid

aceous

period

ACROCANTHOSAURUS

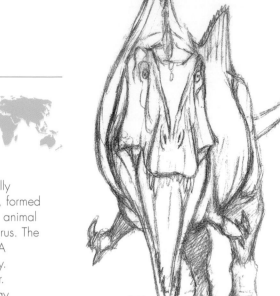

Description: Large theropod
Length: 26–40 ft (8–12m) nose to tail

Distinguishing Features: This large, brightly colored, slow-moving, and exceptionally unintelligent theropod subsists exclusively on carrion. The distinctive sail on its back, formed from extended neck vertebrae, suggests a relationship with Spinosaurus. In fact this animal is much more closely related to Allosaurus or to the "land shark" Carcharodontosaurus. The arms are relatively short and are armed with a pronounced claw on the first digit. A carrion-feeding habit gives these creatures a powerful odor of rottenness and decay. They have a highly developed sense of smell, but their vision and hearing are poor. They are drawn from many miles to rotting carcasses and also to mates—which may of course be found in the same places. These normally solitary animals squabble over carrion, mate in a desultory and cumbersome way, and then leave the scene. Unlike most other terrestrial dinosaurs, Acrocanthosaurus bear live young which are nurtured in a placenta-like arrangement. The offspring are remarkably precocious and run off as soon as they are born. This behavior has a selective advantage because stillborn or slow-moving young are immediately eaten by their mother.

Acrocanthosaurus in frontal exploratory posture. The open jaws do not denote a threat display—the animal is simply sniffing the air, using the Jacobson's Organ at the back of the mouth

Habit and Habitat: Were it not for an extraordinary quirk of dinosaur biology, these animals would be a uniform dull gray. Their normal bright, varied pattern of red and green blotches and stripes is the source of the animals' odor, described variously as "slaughterhouse waste left too long in the sun" and "blocked drains clogged with rotten eggs." Their preference for a diet of rotting meat has exposed them to infection by a variety of bacteria, several of which have developed a close association with the scavenger and have set up colonies in the skin, particularly in the face, back, and sail, causing the unique coloring pattern. The particular bacterial aroma differs from one individual to the next, each animal having its own scent. These differences may influence the otherwise haphazard business of mate choice and perhaps even speciation. Some researchers suspect that bacterial infection is necessary for proper development of the extended vertebrae and the sail, in which the bulk of bacterial colonies are seen, and that animals that remain uninfected are unable to reach sexual maturity and reproduce.

Detail of hand to show prominent claw on first digit

Acrocanthosaurus and smaller theropods scavenge a well-rotted carcass of the sauropod Pelorosaurus that has died during the migration of the herd, seen in the background

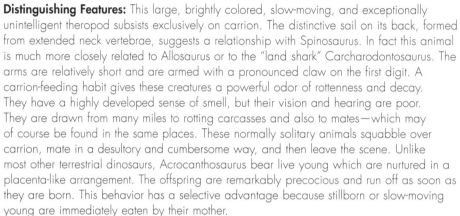

	Triassic		Jurassic		Cretaceous
	245m	208m		146m	65m
Saurischia			Allosaurs		
Theropods					

The footprint of Acrocanthosaurus is of the typical three-toed theropod type

Lateral view of head and neck to show the extended neural spines in the neck vertebrae

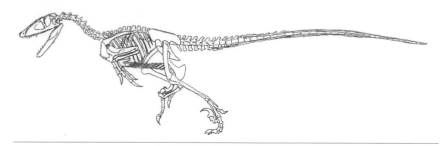

DEINONYCHUS

Description: Pack-hunting theropod
Length: 10 ft (3m) nose to tail

Distinguishing Features: This animal has a large claw on the second toe, relatively long arms, and jaws with wide gape. Males and females are similar in size, though females may be marginally larger. Males are distinguished by prominent display characteristics, including a feathered headcrest. Their color varies from light tan through chocolate brown to deep gray, and can be mottled; many individuals have lighter underparts and small, darker spots on the head and forequarters. The hindquarters and tail may be striped or ringed. The male color is more vivid during the spring breeding season, with prominent roseate wattles and cloacal area, a headcrest that is variable in color, and fringes of black-barred white display feathers on the forearms. Deinonychus builds nests on the ground on the fringes of ornithopod nesting grounds. The chicks are mottled or striped, with a downy pelage and arm and tail feathers that disappear at maturity.

Habit and Habitat: This creature lives in lightly wooded areas to open floodplains, where it follows the herds of ornithopods such as Iguanodon and Tenontosaurus and occasional, rare sauropods. It attacks in groups of between three and six animals, ambushing prey with startling acceleration and then giving chase where necessary, tiring the prey with repeated slashes from its long toe claw. Deinonychus may climb trees in search of lizards and other small vertebrate prey, but avoids water. This dinosaur is extremely dangerous and is best observed upwind through binoculars from an armored truck.

Both parents incubate the large cream-to-teal-blue eggs until they hatch up to 28 days after laying

Juvenile male Deinonychus leaps to the attack

Detail of hind leg showing sickle claw on second toe held clear of the ground

Three views of forearm exhibiting unique hand anatomy and birdlike sideways flexure of the wrist

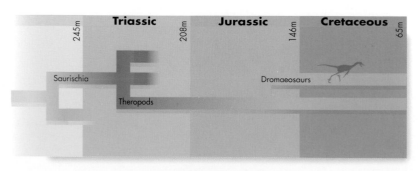

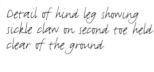

	Triassic		Jurassic		Cretaceous	
245m		208m		146m		65m
Saurischia				Dromaeosaurs		
	Theropods					

Contrast male's
display plumage
(behind) with
the bald,
heavier-set
features
of the
female in
the foreground

Males squabble over a
Tenontosaurus carcass
while a female looks on

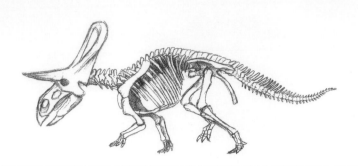

ZUNICERATOPS

Description: Small ceratopsian
Length: 8.2–18 ft (2.5–4m) nose to tail

Distinguishing Features: This small ceratopsian has a very long snout armored with a distinctive bony ridge that runs downward to be continuous with the upper beak. The cheek bones are also prominent and extend sideways, terminating in small horns. It has prominent brow horns (being the earliest known ceratopsian with this feature) and a very large, ornate frill. The "frame" and central spine of the frill are yellowish green with prominent green scutes and accessory horns at the lower, hindmost corners. The central, unsupported regions are outlined in bright orange, particularly in rutting males during the breeding season (see main illustration). The rest of the body is bulky and stocky, with greenish, armored upper regions and yellowish, less well-armored flanks, legs, belly, and tail. These animals form large herds (between 50 and 100 individuals) in which a small number of males compete for females. However, no single male ever has exclusive access to all the females; this means that males fight sporadically for dominance throughout the year, with rutting especially intense in the spring. Females nest communally, creating gigantic midden-heaps of sand and rotting vegetation in which they lay their eggs together.

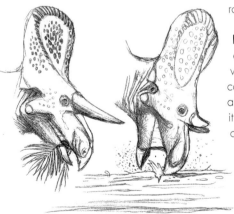

Habit and Habitat: Zuniceratops prefers a wooded environment, swamp forests to semiwooded parkland, where it can browse on shrubs, low-growing trees, and cones, and strip bark from fallen logs in search of insects and grubs. It will occasionally forage for carrion. However, it often roams further afield, and herds tend to migrate to open semidesert areas for the rutting season. Zuniceratops is prey for a number of theropods, which it combats with great ferocity. A lone animal is usually no match for a large theropod or a pair, but if cornered as a herd, Zuniceratops form a defensive phalanx from which some individuals can charge down, attacking the theropods and forcing them to flee.

Zuniceratops, cropping a leaf (left) and stripping bark from a fallen log (right)

Zuniceratops scavenging a theropod carcass

Comparative studies of ceratopsian heads

Styracosa

Centrosaurus

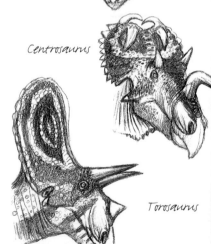

Torosaurus

A young Zuniceratops hatches from atop the midden-heap nest characteristic of this dinosaur

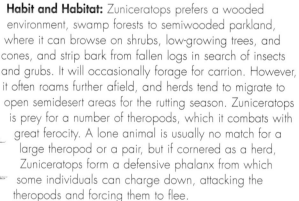

	Triassic		Jurassic		Cretaceous	
245m		208m		146m		65m

Marginocephalians

Ornithischia

Ceratopsians

Raking motion of jaws strips tender
leaves and shoots from tough stems

some species
exhibit the neck
spines embedded in a
double sail—these may be brightly colored as a sexual
ornament during the breeding season

AMARGASAURUS

Description: Medium-sized sauropod
Vital Statistics: 35 ft (10m) nose to tail

Distinguishing Features: This dinosaur cannot be confused with any other animal because of the prominent double row of long, thick spines along its neck and back. In some Amargasaurus species, the spines may be joined by webbed tissue into a "sail" (see sketch left). It has relatively slender, pillarlike legs and a small, boxlike head. It is brown to red on the legs, flanks, and underparts, and gray to black on the spines and associated webbing. Females are 5–10 percent larger than males but duller in color and with shorter spines. Chicks are highly precocial, hatching without spines, which grow to the full length only at sexual maturity.

Habit and Habitat: Amargasaurus lives in small groups in upland gallery forest, grading to larger groups in open lowland during the annual rutting and nesting season. Males use neck-spines for display and occasional combat to establish temporary harem privileges over a group of females. Most of the year, the animals live in female-dominated family groups with loosely associated bachelor groups or solitary males. Generally herbivorous, preferring cones and woody tissue, they take an occasional invertebrate, carrion, or bones. They are occasionally seen in mixed herds with other sauropods (particularly diverse in South America at this time) and small theropods called abelisaurs, distinctive predators of the southern continents.

Peglike tooth

Creamy white
eggs, each slightly
larger than a grapefruit,
have distinctive, large pores

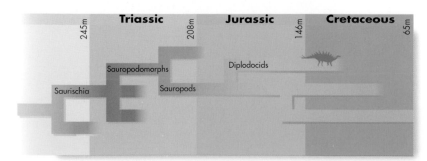

Triassic | Jurassic | Cretaceous

245m | 208m | 146m | 65m

Sauropodomorphs

Diplodocids

Saurischia

Sauropods

Foot of
Armargasaurus

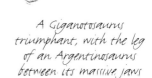

GIGANOTOSAURUS

Description: Large theropod
Length: 43–49 ft (13–15m) nose to tail

Distinguishing Features: One of the largest predators of all time, this ferocious creature looks superficially like a tyrannosaurid but is distinguished by elaborate facial crests and an armor of bony scutes, the three (rather than two) digits on its hands, and the fact that it has four toes (not three), as in later theropods. Males and females are very similar, both with emerald green heads and flanks grading into gray-brown underparts. Giganotosaurus is larger than the Late Cretaceous Tyrannosaurus and comparable in size to the related African Carcharodontosaurus—both members of a group of Mid-Cretaceous theropods related to Allosaurus that evolved as specialist hunters of very large sauropods. Small family bands of Giganotosaurus, usually dominated by a large male, follow migrating sauropod herds and nest on the fringes of sauropod colonies. The male inseminates a number of females in his harem, each of which raises two or three chicks. These offspring are altricial—that is, they are relatively undeveloped on hatching. New blood comes into the group in the form of lone bachelor males who attempt to unseat a dominant male and occasionally succeed. Most of the hunting is done by the females.

Habit and Habitat:

These specialist hunters have the size and muscle needed to attack the largest titanosaurid sauropods. Packs of Giganotosaurus follow a number of sauropod species but are particularly associated with Argentinosaurus. At between 115 and 150 ft (35 and 45m) long, Argentinosaurus is probably the largest land animal that has ever existed. Three or four Giganotosaurus working together can corner a small Argentinosaurus individual and use stealth and cunning rather than speed to subdue it. However, a single Giganotosaurus—even a huge, terrifying alpha male—is no match for a mature Argentinosaurus, which can crush a less capable predator to death beneath its forelimbs, or stun it with a flick of its whiplike tail.

A Giganotosaurus triumphant, with the leg of an Argentinosaurus between its massive jaws

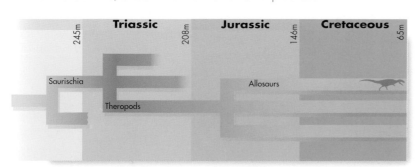

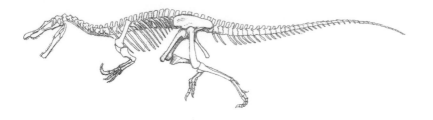

BARYONYX

Description: Large semiaquatic theropod
Length: 30–40 ft (9–12m) nose to tail

Distinguishing Features: This animal has a distinctive, long, slender, crocodilelike skull similar to that of its relatives Suchomimus and Spinosaurus. Note the enormous 12-in (30-cm) claw on the first digit of the hand, and contrast with the claws of distantly related theropods such as Deinonychus, which are on the feet. Note also the "fish-trap" notch in the upper row of teeth toward the tip of the snout. The male (shown in the main illustration) is brightly colored, with bright blue and mauve horizontal stripes, a brown snout, and a small, red crest on the top of the head. The female is slightly larger than the male but duller in color and without the red head crest. These animals are usually solitary and occupy large territories or fishing grounds. As the spring rut approaches, territorial borders "dissolve" as males and females seek out one another and spectacular gatherings of 20 to 30 animals take place in which males compete for females. Animals of both sexes build huge dams of vegetation on islets or rocks in watercourses, in which they lay two or three eggs and take turns incubating the clutch. Baryonyx living close to the sea build nests on raised ground, often within pterosaur rookeries. The young are precocial and are clothed in cryptically colored brown, downy feathers which are shed when they are ready to leave the nest.

Habit and Habitat: Never found far from water, each animal defends a stretch of river or beach as its own. It forages by wading, trapping fishes such as Lepidotes in its jaws. Although a famous fisheater—the heron of the Cretaceous—it catches any aquatic life, including turtles, placodonts, plesiosaurs, and crocodiles. It also takes carrion, including the carcasses of other dinosaurs. Occasional strandings of giant pliosaurs may bring Baryonyx individuals together from far and wide, leading to fierce squabbles over scavenging rights.

A pair of Baryonyx perform a pincer attack on two juvenile Iguanodon

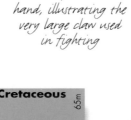

Detail of arm and hand, illustrating the very large claw used in fighting

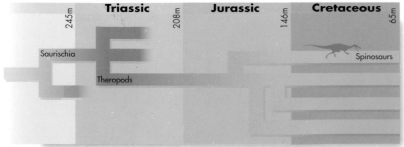

	Triassic		Jurassic		Cretaceous	
	245m		208m		146m	65m
Saurischia						Spinosaurs
	Theropods					

side view of Baryonyx head showing how it uses its jaws as a fish trap

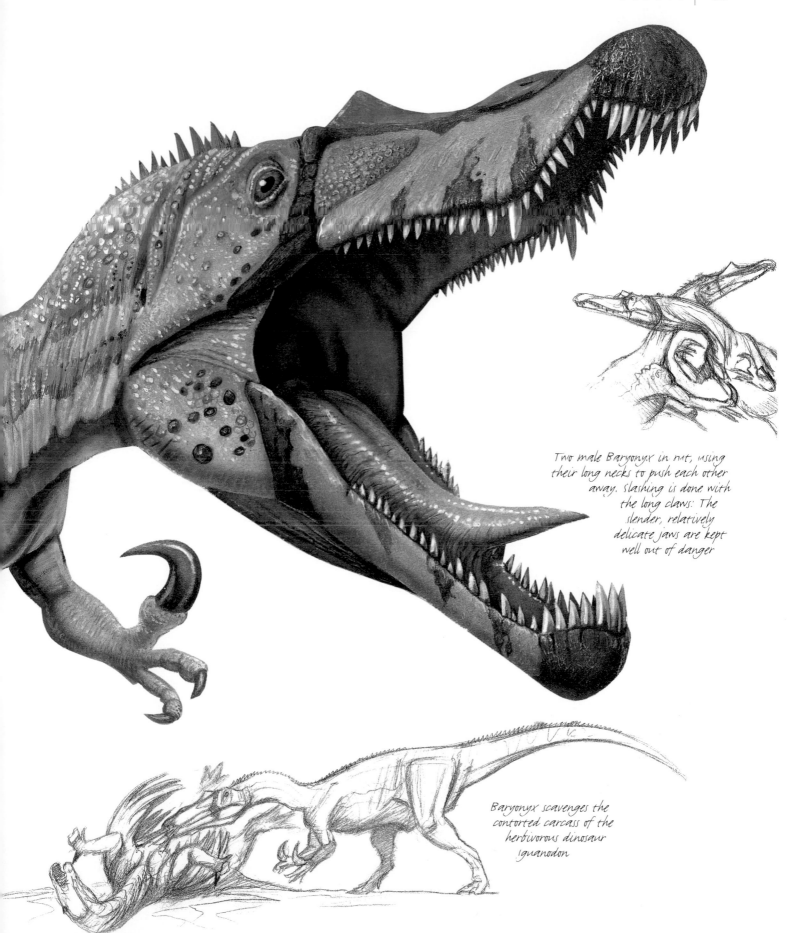

Two male Baryonyx in rut, using their long necks to push each other away. Slashing is done with the long claws. The slender, relatively delicate jaws are kept well out of danger

Baryonyx scavenges the contorted carcass of the herbivorous dinosaur Iguanodon

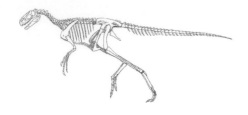

Head of Eotyrannus adult. Note the deep, rather square profile

Head of Eotyrannus chick

EOTYRANNUS

Description: Small to medium-sized theropod
Length: 13–15 ft (4–5m) nose to tail

Distinguishing Features: This lightly built theropod possesses an extraordinarily large, deep-jawed head that seems out of proportion to the otherwise gracile, rangy form of the animal—an oddity that turns out to be a key feature of this dinosaur's ecology. The distinctive head also provides a clue to its relationships because Eotyrannus is one of the earliest known members of the group that later produced such giants as Tyrannosaurus and Tarbosaurus. The bare skin of the head and legs are golden in color, fading to gray. The body and neck are clothed in a pelage of fibrous feathers, often in a leopard-spot pattern of brown or black markings on a white or yellow background. Males and

females produce a clutch of five or six eggs, and the chicks are invariably of the same sex. They are raised together and often remain in a bachelor-sibling pack throughout young adulthood until they themselves mate. The chicks have a very thick plumage, including long-veined contour feathers on the fringes of the arms, as in the subadult shown in the main illustration. All but extremely old animals retain a feathery coat. In tyrannosaurs the tendency is toward a reduction in feathering until it is found only in young chicks in large Late Cretaceous forms. Extrapolating this tendency in reverse has led some to speculate that tyrannosaurs descended from creatures with a full complement of feathers throughout life—in other words, early birds.

Habit and Habitat: A highly intelligent pursuit predator, Eotyrannus forms packs that live on the fringes of the huge herds of Hypsilophodon and other ornithopods such as Iguanodon found in mid-Cretaceous Europe. The association with Hypsilophodon is of additional interest in that the symbiotic alga responsible for the secondary sexual characteristics of the ornithopods fulfills a similar function in Eotyrannus. The predator can become infected only by consuming live, infected Hypsilophodon—especially fierce, mature males. Once inside the tyrannosaur, the algae distort the animal's growth pattern, deepening and enlarging the head. Big heads are favored by both sexes, promoting the spread of a trait which, incidentally, is associated with the ability to capture large Hypsilophodon males successfully. Late Cretaceous tyrannosaurs have large heads without the benefit of such symbioses. It is possible that over millions of years of tyrannosaur evolution—between the Mid Cretaceous and the Late Cretaceous—the genome of the alga introgressed into the larger genome of the later tyrannosaurs. This is just one of the intriguing questions that the ongoing Tyrannosaurus Rex Genome Project (TREGPO) is seeking to answer.

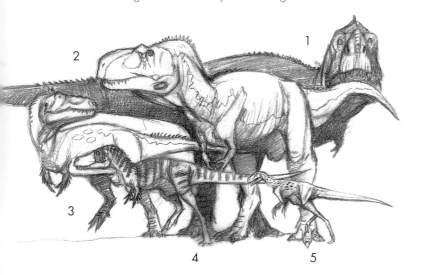

The tyrannosaur family is very diverse, and includes
1. Tyrannosaurus rex; 2. Daspletosaurus; 3. Alioramus;
4. Nanotyrannus; and 5. Eotyrannus

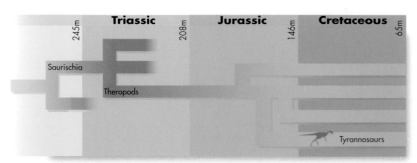

Hand of Eotyrannus, showing the long plumes of the forearm

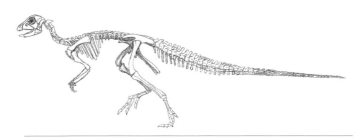

HYPSILOPHODON

Classification: Small ornithopod dinosaur
Length: 5–7.5 ft (1.5–2.3m) nose to tail

Distinguishing Features: This small, brightly colored ornithopod is very common. Males and females appear identical except during the mating season. At all times of the year both sexes have bright, almost fluorescent green heads and flanks, and the tail is barred with conspicuous green stripes. The effect of this barring is to confuse lone predators, especially when (as is usual) ornithopods are gathered together in large numbers. During the spring mating season, males bear a small head crest of blue plumage, and their large eyes are a deep, ominous red (at other times of the year they are yellow, as are the eyes of females at all times). This red color is believed to be caused by a metabolic by-product of a dinoflagellate (algal) symbiont released in response to male sex hormones. This same microscopic partner is responsible for the vividness of the animals' green color. Females prefer males with redder eyes, a characteristic that ensures the propagation of both dinosaur and alga. The same alga also features in the life of Eotyrannus, a major predator of Hypsilophodon. Like many ornithopods, Hypsilophodon lives in very large groups of more-or-less related animals. Mating is random and females nest communally; the nests are then guarded by platoons of patrolling males. Inbreeding is prevented by the sheer size of the groups. Herds more than 1000 strong have been recorded, and there are anecdotes about the existence of even larger ones. Juveniles are drab gray until they become infected with the symbiont. "Quasi-albino" animals, resistant to infection, have been seen, but they are extremely rare. Unlike those of larger ornithopods, the digits of the hands and feet have not been consolidated into mittenlike pads. The hand has all five digits, but the fourth and fifth are stumps. There are four clawed toes on each foot.

A group of Hypsilophodon fleeing from a Baryonyx. The flight pattern is not random, but carefully "choreographed" to cause maximum confusion—an example of Hypsilophodon swarm intelligence

Male Hypsilophodon in threat posture, showing beak teeth

Habit and Habitat: These versatile animals live in most habitats, from low-lying swamp forests to arid uplands, though they prefer lightly wooded plains and hillsides. Being small, they can roost in trees, sometimes in large numbers. These "goats" of the Cretaceous can and do live on virtually any type of food, which they process with their protruberant teeth and sharp beaks. A disconcerting feature of these dinosaurs, reported by several naturalists, is that although the intelligence of an individual Hypsilophodon is only average (and probably less than that of a dromaeosaurid theropod of equivalent mass), the swarm intelligence of many animals gathered together seems uncanny, as evidenced by their lightning-fast, coordinated responses to a threat or to the presence of conspecifics or prey.

Head in lateral view

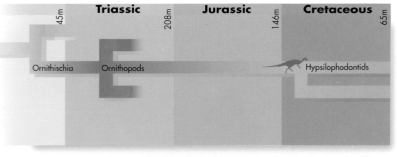

	Triassic		Jurassic		Cretaceous	
45m		208m		146m		65m
Ornithischia	Ornithopods				Hypsilophodontids	

Overleaf: A raiding Eotyrannus surprises a herd of Hypsilophodon

IGUANODON

Description: Large ornithopod
Length: 20–33 ft (6–10m) nose to tail

Distinguishing Features: This large ornithopod is charcoal gray on the upper parts grading into creamy-white on the flanks and underparts and has colorful ornamental head crests and throat pouches. In *Iguanodon mantelli*, shown here, the crests and throat pouches are bright red. Note the large, distinctive thumb spike and the prominent, horny blue-gray beak. Like most ornithopods, Iguanodon is very gregarious and is invariably found in large herds numbering several hundred animals and occasionally up to 2000. Herd size provides a defense against hunters such as Eotyrannus and Deinonychus. Mating displays are noisy and violent as males inflate their throat pouches, hoot at one another, and charge, flailing their occasionally lethal thumb spikes. Males mate with as many females as they can and females incubate large clutches in communal rookeries. Paternity is not random, however—sperm from different males compete for access to eggs in the female reproductive tract. This phenomenon, known as "sperm competition," may have unusual results, as some sperm are highly toxic to females. These effects are reflected in the pattern of Iguanodon life history—successful males reach sexual maturity at five years and can live for 60 or 70 years. Females, however, manage less than half that life span although—paradoxically—the shortest-lived females tend to lay the most eggs.

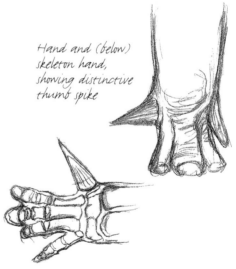

Hand and (below) skeleton hand, showing distinctive thumb spike

Habit and Habitat: Iguanodon is the Model-T Ford of the Cretaceous. Common and widespread from moderately dense forests to open plains, the genus has a very long time range, and any visitor to the Early Cretaceous or Mid-Cretaceous of Europe or North America is virtually guaranteed a sighting of one or more of the 27 recorded species of Iguanodon. The diet of *I. mantelli* is typical for the genus. It feeds on tough, fibrous vegetation broken down in a crop, as well as by bacteria in a complex, ruminantlike stomach. This diet is augmented by small animals, carrion, eggs, and chicks—including those of its own species.

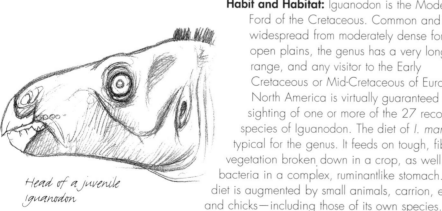

Head of a juvenile Iguanodon

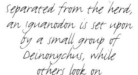

separated from the herd, an Iguanodon is set upon by a small group of Deinonychus, while others look on

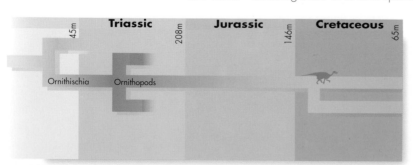

	Triassic		Jurassic		Cretaceous	
245m		208m		146m		65m
Ornithischia	Ornithopods					

SCIPIONYX

Description: Small theropod
Length: 3–5 ft (1–1.5m) nose to tail

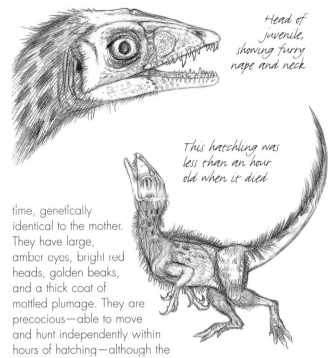

Head of juvenile, showing furry nape and neck

This hatchling was less than an hour old when it died

Distinguishing Features: Typical of the many kinds of small theropod found everywhere throughout the Mesozoic, *Scipionyx samniticus* (the only member of the genus) is common locally in Italy in a restricted slice of the Mid Cretaceous. What is unusual about this animal is that, perhaps uniquely among dinosaurs, it is an obligate parthenogen. That is to say, females reproduce with no intervention from males, which are unknown. Adults (that is, adult females) are dull, reddish brown with sparse, mottled plumage on the neck and trunk. Their snouts, however, are a rich golden color, contrasting with their large, deep-red eyes. Small groups of adults nest on muddy, reedy banks close to streams at any time throughout the year, each female producing clutches of precisely eight eggs which hatch in just five days—a record time for any dinosaur. The offspring are octuplets, products of the fission of a single egg—and, at the same time, genetically identical to the mother. They have large, amber eyes, bright red heads, golden beaks, and a thick coat of mottled plumage. They are precocious—able to move and hunt independently within hours of hatching—although the mother (and other females in the group) bring small prey items for a day or two, as shown in the main illustration. The young grow rapidly and are capable of producing their own offspring within a week of hatching but rarely live beyond two months. The small size of the adult dinosaur, together with features such as its large eyes, suggests paedomorphosis—a tendency to become sexually mature while still in the juvenile state. This of course accentuates the superfast life-cycle of this strange creature.

Habit and Habitat: Scipionyx inhabits densely to sparsely wooded water margins, especially those of river bends and shallow lakes with abundant, ephemeral prey such as frog spawn and insect larvae, as well as fish, small mammals, and general carrion and detritus. However, Scipionyx also nest and breed close to the carcasses of large dinosaurs where they and their offspring feed on flesh and on the larvae of carrion-loving insects. Concentration on such a short-lived food supply is thought to have spurred the development of parthenogenesis, which is an efficient way for a species to exploit rich but short-term resources. A single Scipionyx female has the potential to produce more than 4000 offspring in a month and a half, although the vast majority of these will succumb to predation. Parthenogenetic species are extremely vulnerable to accumulated genetic mutations, which may explain the uniqueness of *S. samniticus*, and its relatively narrow temporal and geographic range.

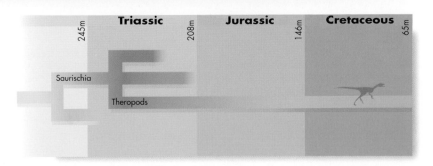

	Triassic		Jurassic		Cretaceous	
245m		208m		146m		65m

Saurischia

Theropods

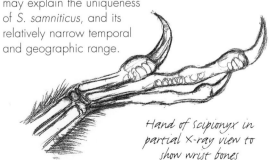

Hand of scipionyx in partial X-ray view to show wrist bones

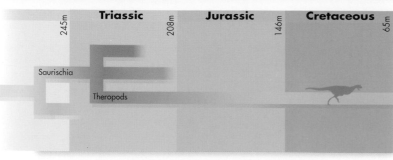

245m	**Triassic**	208m	**Jurassic**	146m	**Cretaceous**	65m

Saurischia

Theropods

Head of Ornithocheirus, a pestilential
pterosaur often found close to
Carcharodontosaurus nesting and kill sites

CARCHARODONTOSAURUS

Description: Large theropod
Length: 26–46 ft (8–14m) nose to tail

Distinguishing Features: One of the largest predators of all time, Carcharodontosaurus is related to Allosaurus and the South American theropod Giganotosaurus. With immense, deep bodies and huge heads, these animals are reminiscent of tyrannosaurids although they are easily distinguished by their larger arms and hands, which have three digits. Males and females are identical in appearance: Greenish on the head, flanks, and tail, grading to red beneath, and with elaborate armor on the head and around the eyes. Males and females pair for life, producing a clutch of two or three eggs a year. The parents share incubation duties and raise the offspring together, teaching them the skills of hunting. The chicks are thickly clad in "camouflage" down of mottled, greenish-brown feathers, which are lost after a few weeks.

Habit and Habitat: Found in locations ranging from lowland marshes and lightly wooded parkland to semidesert scrub, Carcharodontosaurus hunts for turtles, crocodiles, ornithopods such as Ouranosaurus, and small- to medium-sized sauropods such as Aegyptosaurus (main picture) and Paralititan. In this behavior it differs from Giganotosaurus, whose habit of hunting very large sauropods has led to a gregarious, pack-hunting lifestyle.

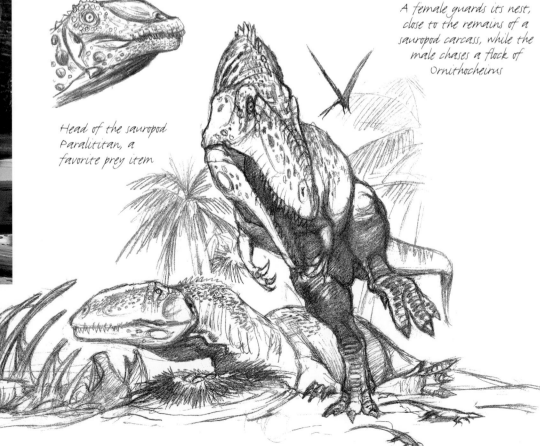

A female guards its nest, close to the remains of a sauropod carcass, while the male chases a flock of Ornithocheirus

Head of the sauropod Paralititan, a favorite prey item

OURANOSAURUS

Description: Sail-backed ornithopod
Length: 20–26 ft (6–8m) nose to tail

Distinguishing Features: The only ornithopod with a sail—skin stretched across extended vertebral spines—Ouranosaurus cannot be mistaken for any other dinosaur. It is mauve to purple, with vertical yellow-brown stripes, especially on the sail, which is topped with a fringe of dense fibers that are more like porcupine quills than the feathers of theropods and are similar to the tail quills of the (unrelated) Psittacosaurus. The head is long, with the snout extended into a horny beak. Wattles on the face can be inflated to augment vocalizations. Like most ornithopods, Ouranosaurus has an elaborate social structure, living in very large migratory herds of up to 200 animals. Animals tend to belong to loosely related clans of 20–30 individuals. Males and females do not form stable pair bonds, and both sexes mate with multiple partners, usually but not exclusively from inside the immediate clan group. Females nest on the ground in huge rookeries, which dominate the landscape and cover several square miles. Each female lays four to six eggs, and females often guard adjacent nests as well as their own. Vigilance and safety in numbers are vital for defense against predators such as Carcharodontosaurus.

Habit and Habitat: This creature lives in open, semiarid scrubland with sparse, low-growing vegetation. As in Spinosaurus, the sail is mainly used as a device to regulate temperature, though it also serves in mating and threat displays to make the animal seem larger than it is. Although it has a superficial resemblance to the hadrosaurs, Ouranosaurus is more closely related to ornithopods such as Iguanodon. Instead of a grinding pavement of teeth, its teeth are arranged in a single row, for slicing and cutting rather than for grinding. Although this dinosaur forages for vegetation, preferring young, tender growth, it derives most of its nourishment from underground. It has a keen sense of smell, hunting for buried invertebrates, fungi, and tender roots which it excavates with its beak and thumb spikes. It also takes small mammals, nesting birds, pterosaurs, and eggs, and eats large quantities of soil—to help grind its varied diet in a gizzard, as well as to boost its mineral intake.

*Ouranosaurus in noisy display
mode, with nasal and throat
pouches fully inflated*

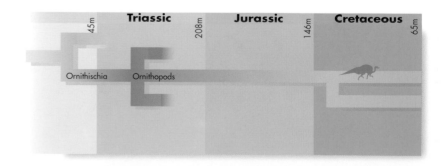

	Triassic	Jurassic	Cretaceous
245m		208m	146m	65m

Ornithischia   Ornithopods

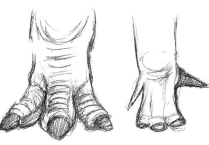

Foot (left)
and hand,
with foot- and
hand-prints

A male and a female scout each other using
their keen sense of smell to determine
relatedness, and therefore suitability as a mate

A herd of Ouranosaurus responds to
the threat of a predator by herding
together, producing a confusing
pattern of ever-shifting stripes

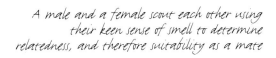

Head of spinosaurus to show gape, and the large collection of very varied teeth

SPINOSAURUS

Description: Very large, sail-backed theropod
Length: 36–66 ft (11–20m) nose to tail

Distinguishing Features: One of the largest theropods that ever existed, Spinosaurus is longer than Tyrannosaurus and its contemporary Carcharodontosaurus but less heavily built. Its great size is emphasized by a prominent "sail" on its back made of skin supported on vertebral spines, which extends outward to 6 ft (2m) or more. As in all spinosaurs, such as the slightly earlier Baryonyx, the head is longer and less deep than in many theropods, with an elongated, crocodilelike snout. The dull gray-brown coloration of this animal is understated in comparison with its dramatic shape. These creatures are usually solitary but form breeding pairs in the spring, incubating a clutch of between one and three eggs in extensive nests in sand dunes. After raising the chicks, each member of the family goes its own way. As with theropods in general, the chicks are feathered, and the down of Spinosaurus chicks is buff to brown, allowing them to merge into their sandy environment if threatened.

A spinosaurus ambushes a plesiosaur resting seal-like on rocks

Habit and Habitat: Spinosaurus lives in the semiarid to arid coastlands of the western Tethys Ocean and the emerging Atlantic. Temperature variation in this environment can be extreme, and the animal's sail, enriched with blood vessels, is used to keep its body temperature within tolerable limits. Like the unrelated Cryolophosaurus, Spinosaurus is a beachcomber; unlike Cryolophosaurus, it is an active hunter as well as a scavenger, taking small, stranded plesiosaurs, pterosaurs, turtles, and large fishes such as coelacanths (see main illustration). It wades some distance out to sea in search of food, using its long jaws and powerful claws to catch its prey.

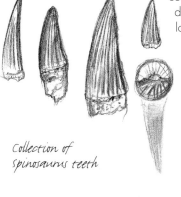

Collection of spinosaurus teeth

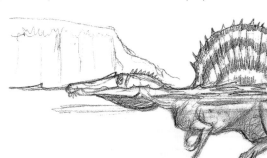

Spinosaurus in swimming posture with snout and sail above the water

	Triassic		Jurassic	Cretaceous	
245m		208m		146m	65m
Saurischia				Spinosaurs	
	Theropods				

The coelacanth: A typical part of the spinosaurus diet

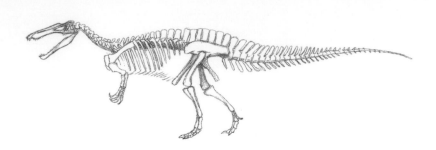

SUCHOMIMUS

Description: Large sea-going theropod
Length: 33–49 ft (10–15m) nose to tail

Distinguishing Features:

This creature can be thought of as a somewhat larger, more aquatic African version of Baryonyx, to which it is closely related. Like Baryonyx, it has very long, slender, crocodilelike jaws, with "fish-trap" notches toward the tips of the snout. It is, however, a more robust creature than its northern relative, and duller in color, with a deeper body, stouter dermal scutes, gray-blue to reddish skin, and truly massive claws. It is also far less sociable than Baryonyx and is invariably found alone. Indeed, no report exists of two of these creatures ever having been seen together. It is thought that males and females usually meet and mate out at sea and that their eggs, like turtles' eggs, are laid on the beaches of remote islands.

Habit and Habitat: Suchomimus inhabits the beaches of North Africa and the adjacent seas. It spends most of its time hunting in the open sea, where it takes turtles, pterosaurs, and fishes. Facial scars and injuries to the face and forelimbs observed on old individuals testify to battles with mosasaurs and plesiosaurs, though no Suchomimus would be able to match the larger pliosaurs in ferocity. The main illustration shows a hazard of shallow-water fishing: The crocodile-mimic is the victim of an attack by the real thing, a 98-ft (30-m) specimen of the ocean-going crocodile, *Chthonosuchus lethei*.

suchomimus as seen head-on

A suchomimus wading in the shallows notices approaching crocodiles

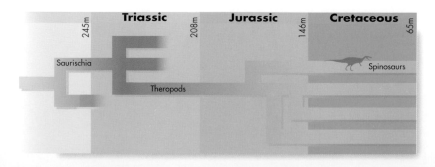

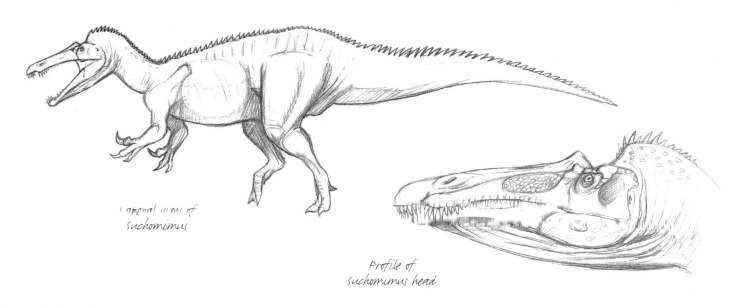

Lateral view of
suchomimus

Profile of
suchomimus head

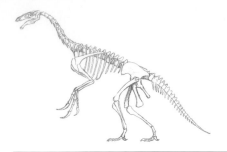

Profile of Beipiaosaurus head, showing the slender, elongated snout and beak

BEIPIAOSAURUS

Description: Small to medium-sized omnivorous theropod
Length: 7–13 ft (2–4m) nose to tail

Distinguishing Features: This peculiar bipedal dinosaur has a very small, birdlike head and a long, graceful neck attached to a huge-bellied, bulky body with very stout legs. The animal is entirely clad in a thick, white pelage except for the lower legs and the belly, which is armored with thick, round scales. The three-fingered hands are armed with enormous claws. Beipiaosaurus is a member of a group of highly modified theropods know as therizinosaurs, which reach their pinnacle in the spectacular Late Cretaceous Therizinosaurus. Note in particular the four-toed foot characteristic of therizinosaurs, believed to have evolved from the three-toed foot of a theropod ancestor. Males and females are all but identical in size and color, and although they are found in fairly large flocks, a male and a female mate for life. The female lays a single clutch of three or four round eggs every two years or so, from which only one chick is expected to hatch. The fledgling often remains close to its parents for many years, sometimes helping to raise siblings in later years before seeking a mate of its own. The group in the main illustration consists of a breeding pair (foreground) and a 3-year-old offspring (background) that has not yet acquired the distinctive blue facial markings seen in adults.

Habit and Habitat: This creature is typically found in dense woodland, where it lives on insects and other invertebrates extracted from bark or rotting logs (see Therizinosaurus), carrion, fungi, and detritus. The variety of birds and birdlike dinosaurs found in northern China in the early Mid-Cretaceous explains the perennial popularity of this locale among naturalists. In addition to Beipiaosaurus being chased by a pack of smaller dromaeosaurid theropod Egovenator (center), the main illustration shows male birds Confuciusornis (top left and right) adorned with long tail-plumes, a nesting pair of feathered oviraptorosaur Caudipteryx (bottom right), and six tiny, long-tailed Sinosauropteryx (foreground).

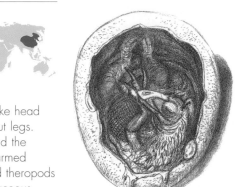

A Beipiaosaurus egg containing an embryo close to full term

A large flock of around ten Beipiaosaurus, looking like gigantic geese

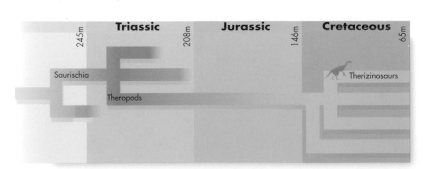

	Triassic	Jurassic	Cretaceous
245m		208m	146m	65m

Saurischia

Theropods

Therizinosaurs

Well-camouflaged eggs are top-shaped—in narrow confines they may roll but not fall

Foot highly adapted to climbing, with long, sharp, curved claws providing maximum grip

Veined feather from the tail, arms, or legs (left), and insulating feather from the body (right)

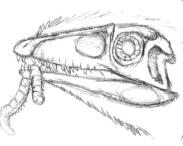

Detail of head with skull exposed. The long snout is adapted for picking caterpillars and other insects off leaves

Only 2 or 3 chicks will survive from a typical clutch of 6 to 8 hatchlings

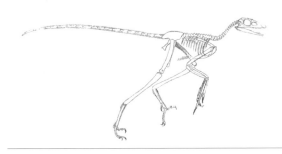

MICRORAPTOR

Description: Small, tree-living feathered theropod
Length: 12 in (30cm) nose to tail

Distinguishing Features: At first this dinosaur is difficult to distinguish at a distance from the bird Confuciusornis: Although both have very long tails, Microraptor is smaller overall and does not fly. Both sexes are thickly furred. The pelage is cryptic—yellow-green to creamy white with black, chocolate, or gray bars, stripes, and spots. The eyes are fringed with a black pelage. The toes on the hind feet are adapted for perching and have long, slender recurved claws. Males and females appear identical. Nests and chicks are hard to observe: The animals nest very high in the canopy and/or in hollow trunks.

Habit and Habitat: This dinosaur inhabits subtropical lowland swamps and rainforests, where it lives in the forest canopy. It can be seen either solitary, in small groups, or in very large, highly vocal assemblies or "parliaments." The high-pitched shrieking of congregating Microraptor is a signature sound of dense lowland forests in East Asia. Its diet is omnivorous: The animals take insects, other small vertebrates such as the mammals Zhangheotherium and Jeholodens, chicks and eggs of other tree-living dinosaurs, and birds. They are partial to the fruits of flowering plants such as Archaeofructis, locally abundant in sheltered river margins. It is suspected that Microraptor might disperse these plants: After eating the fruits, it scatters the seeds in its droppings. Other dinosaurs in this habitat include theropods such as Sinosauropteryx, Sinornithosaurus, Caudipteryx, and Protarchaeopteryx. All are feathered, and although some have a perching habit, none are as small as Microraptor.

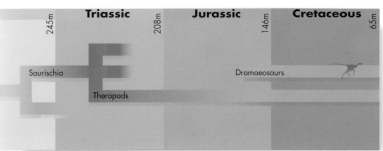

	Triassic	Jurassic	Cretaceous
245m		208m	146m	65m

Saurischia

Dromaeosaurs

Theropods

PSITTACOSAURUS

Description: Primitive bipedal ceratopsian
Length: 3–8 ft (80cm–2.5m)

Distinguishing Features: Easily mistaken for a rather heavy hypsilophodontid, this unusual creature is in fact a primitive offshoot of the stock that eventually gave rise to dinosaurs such as Protoceratops, Zuniceratops, and Triceratops. Unlike these later dinosaurs, Psittacosaurus is mainly bipedal and surprisingly fleet-footed considering its deep belly and rotund appearance. The small head has a distinctively round shape, with a short, deep parrotlike beak (hence the animal's name) and horns emerging from the cheek region. Particularly noticeable is the row of long, quill-like extrusions emerging from the upper margin of the tail. Each quill is tipped with microscopically small but very sharp, recurved points and small reservoirs of venom which explode on contact. Males and females are very similar in appearance and are brownish red. They tend to be solitary, mating whenever they meet in the dark forests where they live. The female lays five or six eggs in shallow scrapes amid tree roots and offspring accompany their mother until adulthood. Chicks are clothed in a mottled pelage and they rely on their ability to escape by running at high speed into the gloomy depths of the forest.

Habit and Habitat: These animals are invariably found in the densest and most tangled parts of tropical forests, where they specialize in eating the fruits and nuts of primitive flowering plants, which during this period are largely confined to the wet tropics. However, like many dinosaurs, they scavenge for carrion and are particularly fond of bones: The fragments make useful gizzard stones. Psittacosaurus is an excellent swimmer and is sometimes found in jungle lakes and streams, dabbling for weedy plants, snails, and other invertebrates. When confronted by predators such as Sinovenator (see illustration overleaf), it swings its tail to good effect. Barbed quills bitten off by predators are hard to dislodge from the mouth before each has shed its load of irritating, toxic phenols and quinones, heated by a chemical reaction connected with removal of the quill from the animal. Shed quills are quickly replaced by regrowth from the base.

Detail of tail-quills

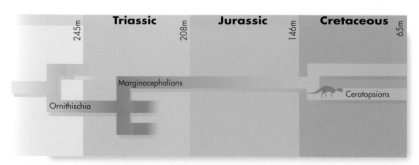

	Triassic	**Jurassic**	**Cretaceous**
	245m	208m	146m	65m

Marginocephalians

Ceratopsians

Ornithischia

Two males face off, displaying their fountain of tail-quills

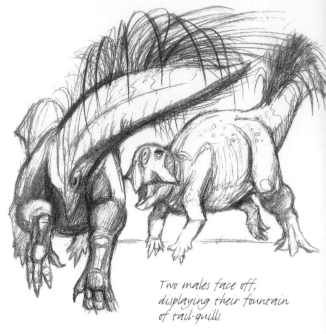

A threatened Psittacosaurus rears up, exposing its gape, and menacingly fanning out its poisonous tail-quills

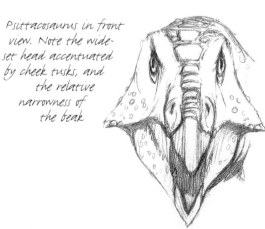

Psittacosaurus in front view. Note the wide-set head accentuated by cheek tusks, and the relative narrowness of the beak

SINOVENATOR

Description: Small theropod
Length: 6–9 ft (1.8–2.6m) nose to tail

Female Sinovenator in pursuit of a male Confuciusornis

Distinguishing Features: This small and very brightly colored theropod is a primitive member of the remarkable group known as the troödontids. Females (see illustration overleaf) have a thick, emerald-green pelage on the face, neck, arms, body and tail, punctuated with bright red, black-edged spots. The legs and undersurfaces are a dull gray. The head has a bright red shock of feathery quills, and the arms bear ranks of stubby green, red, and black feathers. The snout is a deep golden color. Males are even more brightly colored, but in very individual patterns, differing from one another as well as from the females. Prehistoric peacocks, these animals are usually completely clad in feathers, and have long, ornate plumes growing from the head, arms, and tail. Males display to females in leks, usually in sunlit clearings in the deep forest, showing an array of colors described by one observer as "Mardi Gras in Rio." The males' plumage makes it impractical for them to hunt, and they are solely responsible for incubating the clutch of six or seven eggs laid by the female, who takes on full responsibility for obtaining food. Hunting is often undertaken by several females working together.

The nest is guarded by the male while his mate hunts for food

Habit and Habitat: Sinovenator inhabits forested country ranging in density from moderately degraded woodland to deep, primary jungle where it hunts herbivorous dinosaurs such as Psittacosaurus and Beipiaosaurus. It also takes birds (Confuciusornis), eggs, and small mammals such as Repenomamus. Like all troödontids, Sinovenator is a highly visible animal because of its large eyes and bright plumage. The individuality of the plumage in males seems to go far beyond the needs of display or of threat, and some observers think that the subtleties of shade and tone may be part of a sophisticated system of interaction, a visual analog of the songs of whales. This is perhaps not surprising given that troödontids are perhaps the most highly intelligent of all dinosaurs.

Unlike the front tooth (left), the back tooth (right) is deeply serrated

Foot, showing sickle claw on raised second toe

small mammals form a regular part of the animal's diet

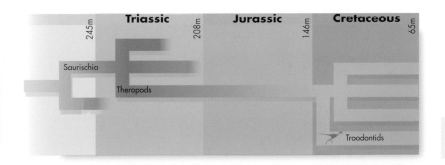

	Triassic		Jurassic		Cretaceous	
245m		208m		146m		65m

Saurischia

Theropods

Troodontids

Overleaf: Psittacosaurus adult and young flee from a Sinovenator attack

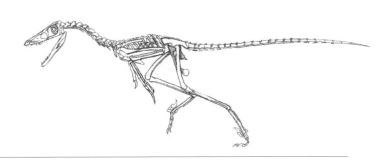

A juvenile, with longer, denser feathers than the more ground-living adult, runs into higher branches to escape a predator

SINORNITHOSAURUS

Description: Small feathered theropod
Length: 1.6–4 ft (50cm–1.2m) nose to tail

Distinguishing Features: This small, largely ground-living theropod is covered with featherlike quills similar to those seen in the contemporary Sinosauropteryx and Microraptor. Adults of both sexes are covered with a uniform pelage of white to pale gray quills up to 1.5 in (40mm) long, except for blue wattles around the eyes, and naked claws and feet. An unusual feature of this dinosaur is the striking plumage of juveniles, as seen in the specimens play-fighting in the main illustration. The coloring is bright red with patches and stripes of bright blue, and the long tail is ringed blue and white. The feathering is also much thicker than in the parents and inclined to be matted. The fringes on the arms tend to stick together like strips of Velcro, forming airfoils that help the animals move quickly into the trees if threatened. Adult plumage is much sparser and less colorful. The reason for the bright color is a mystery, but it may be related to sibling conflict, an evolutionary phenomenon in which chicks compete with one another for their parents attention and the food that accompanies it.

Habit and Habitat: Largely nocturnal, these animals inhabit dense woodland to semiopen parkland where they forage for small mammals such as Zhangheotherium as well as invertebrates. Males and females pair for life and build nests in noisy rookeries in the lower branches of trees—often the same trees whose upper branches are colonized by Microraptor. A rookery may comprise 80–100 animals from several interrelated generations. Unfledged juveniles and young adults often stay at the parental nest to help raise younger siblings and cousins.

juvenile with arms outspread to show structure of hand and general disposition of feathers

Adult head showing teeth

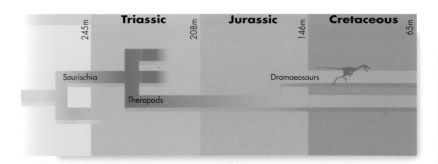

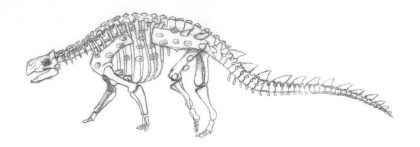

MINMI

Description: Small armored dinosaur
Length: 5–10 ft (1.5–3m) nose to tail

Distinguishing Features: Lightly built for an ankylosaur, Minmi has relatively slender limbs and light armor. The reddish-brown skin has a pebbly texture and is adorned with many moderately large, off-white scutes on the head, back, hindquarters, and tail where they are particularly prominent. When attacked or threatened, the animal will dig itself into the ground within seconds, leaving only its knobbly back exposed. The animals are generally solitary and are never found in groups of more than two or three. Males and females are identical to all but genetic inspection. Mating is brief and can occur at any time of the year. A clutch of six to eight eggs is buried in sand and abandoned.

Habit and Habitat: This extraordinarily tough, resourceful creature is found in a variety of habitats from lush river bottoms—where it feeds on seeds, leaves, and invertebrates—to dry, sandy desert. Roaming for hundreds or thousands of miles, Minmi can go without water for months, storing fat in subcutaneous deposits. *In extremis* it goes into a hibernation mode, digging a burrow in which it buries itself alive, and stays in suspended animation for many months.

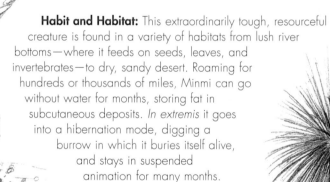

Minmi digs
itself into the
ground

Newly hatched Minmi digs
its way out of a buried nest

selection from a Minmi
menu: centipede, bone, and
tubers and seeds from an
early flowering plant

Now completely buried,
it is hard to know when
dinosaur ends and
desert floor begins

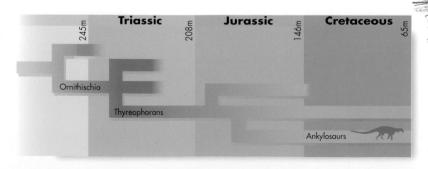

	Triassic	Jurassic	Cretaceous
245m		208m	146m	65m

Ornithischia

Thyreophorans

Ankylosaurs

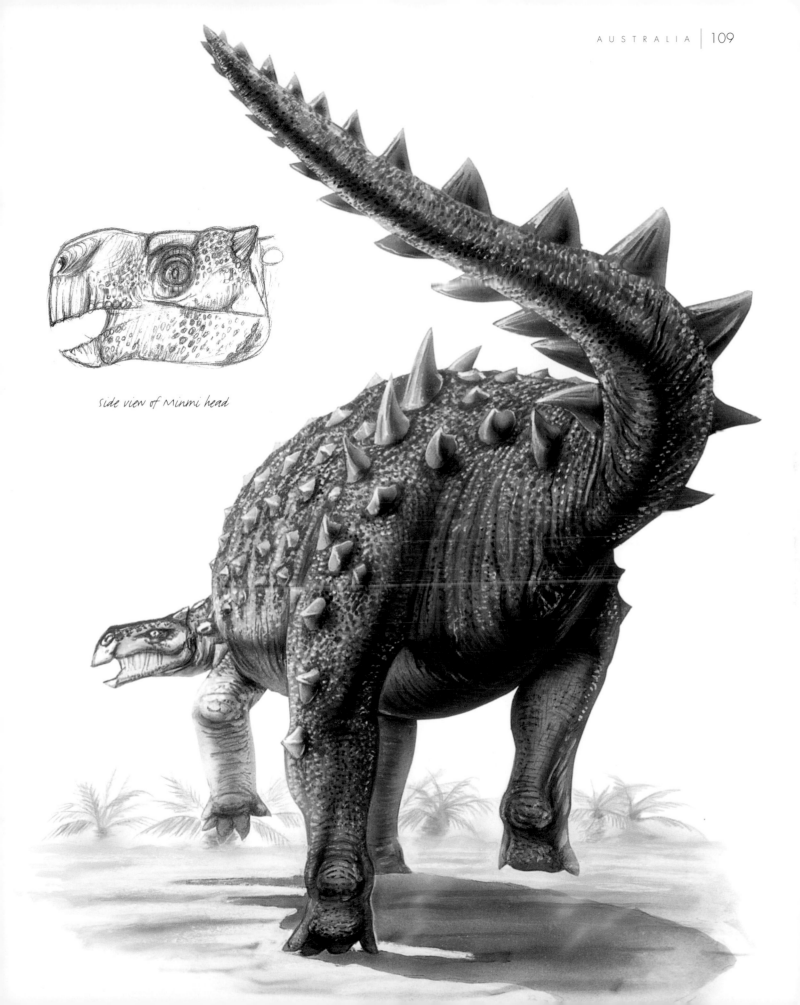

side view of Minmi head

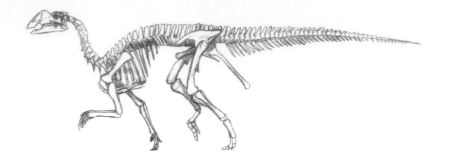

MUTTABURRASAURUS

Description: Medium-sized ornithopod
Length: 20–26 ft (6–8m) nose to tail

Distinguishing Features: Closely related to its near contemporary Iguanodon, Muttaburrasaurus is a common, fairly typical ornithopod. Males and females are generally gray, with pinkish underparts. These animals are distinguished by a prominent nasal crest bearing red nasal sacs, which are inflated during vocalization, and a tough, horny beak. A spiny crest runs along the backbone. Males and females are approximately equal in size, though males have larger nasal crests and sacs. These are used in noisy, violent contests during the breeding season, when males compete for females in a frenzy of hooting and bruising shoulder-jostling contests. The contender that inflicts a ritual bite on the back of its opponent usually wins the battle. Like all ornithopods, Muttaburrasaurus has a busy social life. These animals are found in large, mixed herds and form no stable family units. Males try to inseminate as many females as they can, and after mating, females lay eggs in extensive, cooperative rookeries.

Habit and Habitat: Found in diverse environments from lush floodplain pastures to moorland and scrub at moderate altitudes, this animal eats a variety of vegetation and virtually anything else it can find, including bones and the eggs and young of other vertebrates. There are reports of Muttaburrasaurus invading pterosaur rookeries and biting the legs off young chicks, a grotesque way of increasing its calcium intake. The snapping beak augments a grinding gizzard and a battery of densely-packed teeth, although there are reports that Muttaburrasaurus, uniquely for ornithopods, replaces its teeth all at once rather than one at a time.

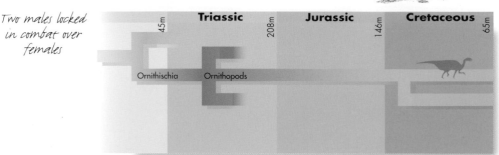

side views of male head, with nasal sacs inflated or deflated

Two males locked in combat over females

	Triassic	Jurassic	Cretaceous
45m	208m	146m	65m

Ornithischia   Ornithopods

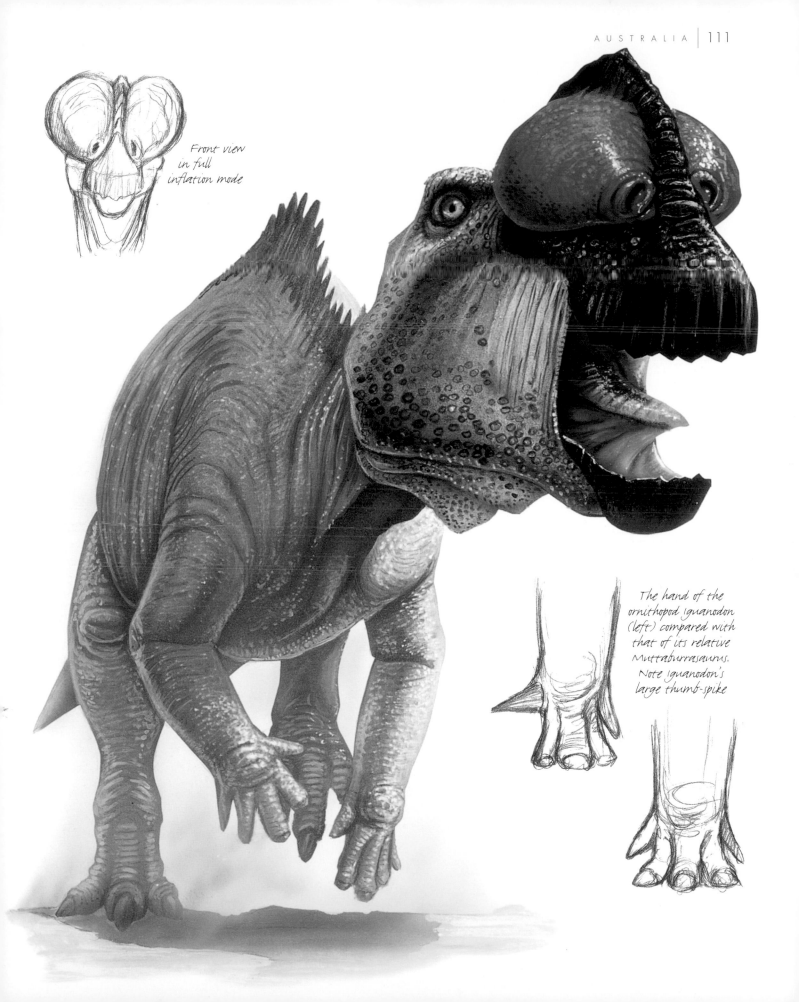

Front view
in full
inflation mode

The hand of the
ornithopod Iguanodon
(left) compared with
that of its relative
Muttaburrasaurus.
Note Iguanodon's
large thumb-spike

The late

Cretac

100 to 65 million years ago

eous

period

EDMONTONIA

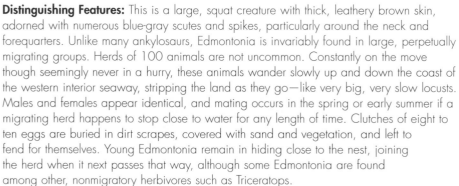

Description: Large armored dinosaur
Length: 20–23 ft (6–7m) nose to tail

Distinguishing Features: This is a large, squat creature with thick, leathery brown skin, adorned with numerous blue-gray scutes and spikes, particularly around the neck and forequarters. Unlike many ankylosaurs, Edmontonia is invariably found in large, perpetually migrating groups. Herds of 100 animals are not uncommon. Constantly on the move though seemingly never in a hurry, these animals wander slowly up and down the coast of the western interior seaway, stripping the land as they go—like very big, very slow locusts. Males and females appear identical, and mating occurs in the spring or early summer if a migrating herd happens to stop close to water for any length of time. Clutches of eight to ten eggs are buried in dirt scrapes, covered with sand and vegetation, and left to fend for themselves. Young Edmontonia remain in hiding close to the nest, joining the herd when it next passes that way, although some Edmontonia are found among other, nonmigratory herbivores such as Triceratops.

Habit and Habitat: One of the most successful and most common of all ankylosaurs, Edmontonia exemplifies the proverbial hardiness of this extraordinary group of animals. Remarkably fierce if roused, they defend the herd against attack from tyrannosaurs and other predators by facing the attacker as a group, swaying from side to side so that their shoulder spikes move with a powerful scything action. This crablike movement is also used by individuals jostling for mates or food. They subsist on anything that can be ingested or sheared off with their powerful beaks—whether vegetation, nuts, roots, invertebrates, eggs, small mammals, young dinosaurs, carrion, or just detritus—provided that it does not hold up their inexorable migration.

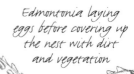

Edmontonia laying eggs before covering up the nest with dirt and vegetation

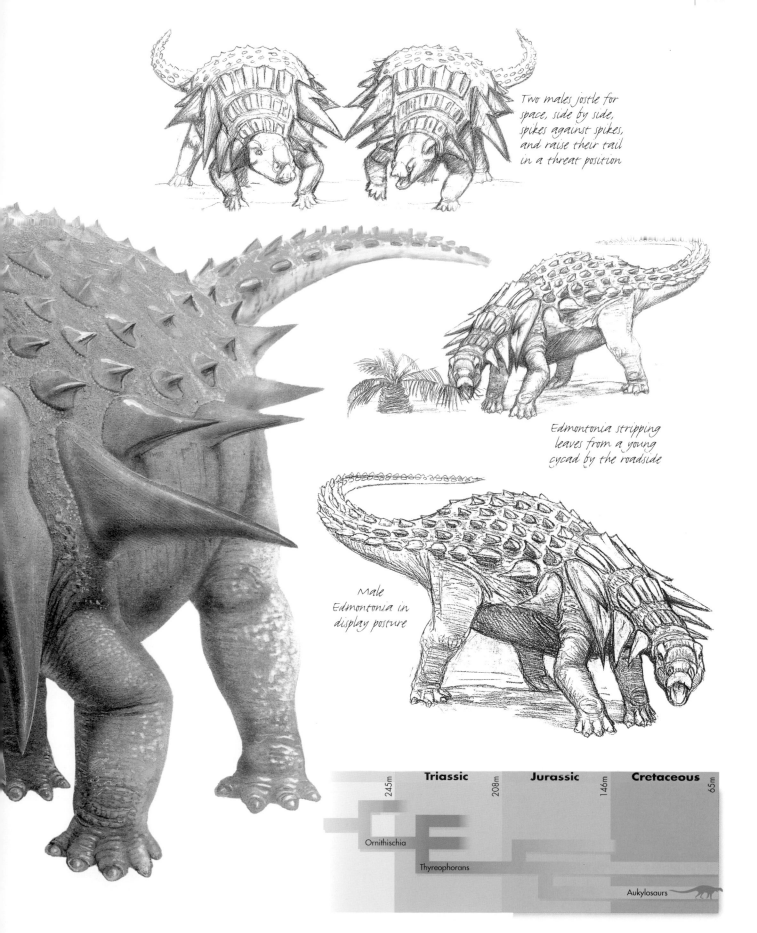

Two males jostle for space, side by side, spikes against spikes, and raise their tail in a threat position

Edmontonia stripping leaves from a young cycad by the roadside

Male Edmontonia in display posture

	Triassic	Jurassic	Cretaceous
245m	208m	146m	65m

Ornithischia

Thyreophorans

Aukylosaurs

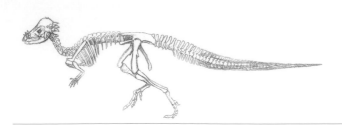

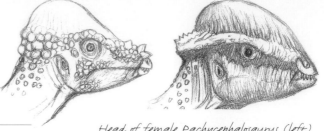

Head of female Pachycephalosaurus (left) and Stegoceras (right). Compare the different head crests of the two species

PACHYCEPHALOSAURUS

Description: Large pachycephalosaur
Length: 16–30 ft (5–9m) nose to tail

Distinguishing Features: The largest of a number of related species of "bone-headed" dinosaurs, Pachycephalosaurus also shows the greatest sexual dimorphism. The male is relatively small (about 20 ft [6m] long) but brightly colored, with an almost iridescent green head, forearms, and flanks grading into a brown tail and hindquarters. The distinctive skull dome, ringed with spikes, may be 12 in (30cm) thick or more. Females are significantly larger than males—up to 30 ft (9m)—but are darker and drabber in color and have few spikes on the head. Social life is polyandrous and violent. A single dominant female can have a harem of as many as 10 or 12 males. Subsidiary females, often related to the dominant "queen," assist in tending the queen's clutches of 50 or more eggs sired by any or all members of the harem. These females lay far fewer eggs, and their young are less likely to reach maturity. Subsidiary females frequently challenge the queen, which results in ferocious contests as each contender tries to crush the barrellike rib cage of its opponent, using the skull dome as a battering ram. Battles take place in the nesting grounds, and there is often considerable collateral damage in the form of trampled eggs and young. Males may also spar for the attention of the queen, with similarly disastrous results.

Habit and Habitat: These animals live in large groups occupying extensive home ranges in open or lightly wooded parkland where they graze on low-lying vegetation. With nowhere to hide from predators, Pachycephalosaurus opts for confrontation, charging all intruders invading the home territory. Even large tyrannosaurs are head-butted in the rib cage, underparts, and legs by bands of males. Adult tyrannosaurs sometimes sustain serious wounds, and for young tyrannosaurs these encounters are often fatal. The belligerence and the social structure of typical Pachycephalosaurus may be adaptations to the presence of gigantic predatory theropods. This creature should be approached only with extreme caution and in a heavily armored vehicle.

Pachycephalosaurus males charge a young Tyrannosaurus rex, temporarily winding it

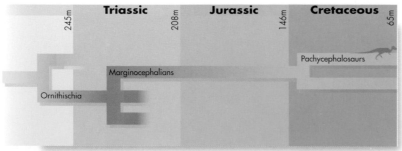

	Triassic	Jurassic	Cretaceous
245m		208m	146m	65m

Pachycephalosaurs

Marginocephalians

Ornithischia

TRICERATOPS

Description: Large / Antipsian
Length: 23–33 ft (7–10m) nose to tail

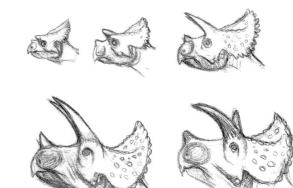

A graded series showing the heads of Triceratops from very young (top left) to fully mature (bottom right)

Distinguishing Features: This is a large, bulky, horned dinosaur with a relatively short, solid neck frill, a pair of long horns over the eyes, a short, forward-curving nose horn, and a prominent, strong beak. Its gray to greenish-black hide is rough and armored with many small, bony bosses. Its neck frill—unlike that in many smaller ceratopsians—is not highly colored or ornamented. Triceratops lives in small herds dominated by large bulls; the spring breeding season is marked by fierce battles for supremacy and access to the harem of females. Each female lays 15–20 eggs in large, round nests walled with hard-packed earth and covered with conifer branches. The animals are not truly migratory, but each herd has its own strictly defined home range, and individuals in one herd rarely cross into the home range of another. This habit—combined with the strict dominance hierarchy—means that many herds of Triceratops are highly inbred. Reports from near the end of the Cretaceous period document large-scale hatching failures and suggest that many herds are dying out.

A herd of Triceratops being stalked by three Tyrannosaurus rex

Massive forefeet are adapted for digging up roots and grubs

Habit and Habitat: This, the largest (and, as it turns out, the last) ceratopsian lives in low-lying coniferous swamp forests to semiwooded parkland where it feeds on woody shrubs and conifer cones. This animal also digs for roots; the beak and nose horn are used to strip bark from rotting logs as it looks for worms and grubs, which it collects on a long, prehensile tongue. The brow horns can be used to shift logs and vegetation to the same ends. This large, well-armored creature has no serious competitors or enemies except for the gigantic theropod Tyrannosaurus, whose large size means that it can cover ground relatively quickly; Triceratops, in contrast, defends itself by standing and fighting. In most cases, Tyrannosaurus gives up if seriously threatened, as even a minor wound from a Triceratops horn could prove fatal.

A female attending a nest

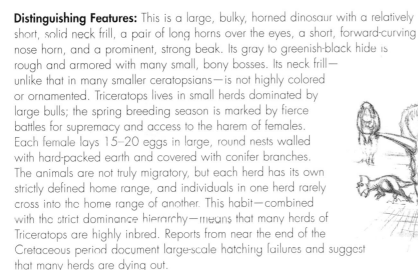

	Triassic		Jurassic		Cretaceous	
245m		208m		146m		65m

Marginocephalians

Ceratopsians

Ornithischia

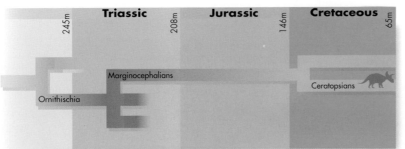

Overleaf: A surprise attack on Triceratops' undefended hindquarters is the predator's best chance of a kill

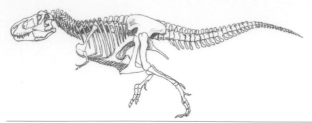

TYRANNOSAURUS

Description: Large theropod
Length: 33–49 ft (10–15m) nose to tail

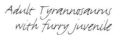

*Adult Tyrannosaurus
with furry juvenile*

Distinguishing Features: This very large, heavily built theropod has a disproportionately large, deep head, and vestigial arms with hands reduced to two small digits. Of the several species, the best known is *Tyrannosaurus rex* (shown here). Males and females are similar in size and appearance, however, females may be marginally larger, and males have more prominent facial ornamentation. They are blue-gray on the dorsal surfaces, grading to deep red or even purple on the flanks, legs, and underparts. Although individuals spend much of their time as solitary hunters, they form loose-knit packs consisting of a dominant male, a harem of two or three females and one or two bachelor males. The bachelors occasionally challenge the alpha male for dominance, and such encounters are very often fatal. Females build huge nests of rotting vegetation and supply carrion for the chicks. The smell of decay from Tyrannosaurus nests can be detected for miles downwind. Chicks are richly clad in black and white feathery down which they shed after a few weeks. At up to 6 tons in weight, *T. rex* is almost certainly the most massive terrestrial carnivore known from any age, even though some earlier theropods such as Giganotosaurus are in fact longer and taller. There are reports that *Tyrannosaurus helcaraxae* (a rare, woolly, hadrosaur-hunter known only from the Late Cretaceous of the north slope of Alaska) is even larger, but this has not been confirmed.

Habit and Habitat: Roaming freely in most lowland habitats from dense woodland to open parkland and floodplains, this creature has little to be afraid of. Although a pursuit predator, it specializes in slow moving, often armored prey—dinosaurs such as Triceratops and Edmontonia—so that its pursuit is somewhat sedate when compared with that of smaller theropods. The unhurried pace is compensated for by armor-piercing teeth—a bone-crushing ability enabled by deep jaws, a short neck, a strong back, and powerful legs and feet. The teeth of *T. rex* can penetrate the bony frill of Triceratops, and stool samples consist largely of crushed bone. But the animal is no hero, and usually backs off if its prey puts up any kind of spirited defense.

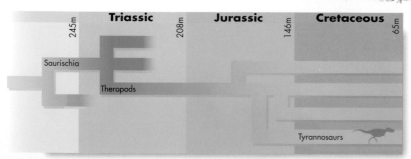

	Triassic		Jurassic		Cretaceous	
	245m		208m		146m	65m

Saurischia

Theropods

Tyrannosaurs

*Tyrannosaurus
showing gape*

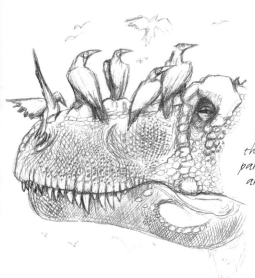

A small flock of birds congregates on the nose of a dozing Tyrannosaurus. Very large theropods depend on birds to remove parasites from their teeth, nostrils, and the soft skin around their eyes

A Tyrannosaurus gloats over the carcass of a fresh kill, the large hadrosaur Anatotitan

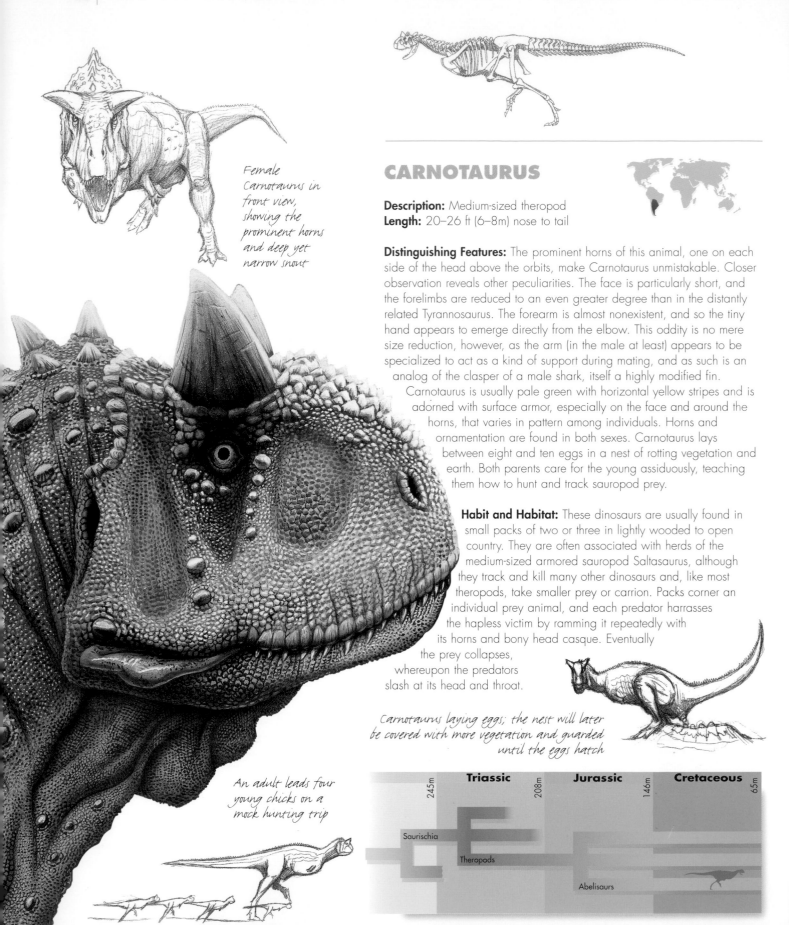

Female Carnotaurus in front view, showing the prominent horns and deep yet narrow snout

CARNOTAURUS

Description: Medium-sized theropod
Length: 20–26 ft (6–8m) nose to tail

Distinguishing Features: The prominent horns of this animal, one on each side of the head above the orbits, make Carnotaurus unmistakable. Closer observation reveals other peculiarities. The face is particularly short, and the forelimbs are reduced to an even greater degree than in the distantly related Tyrannosaurus. The forearm is almost nonexistent, and so the tiny hand appears to emerge directly from the elbow. This oddity is no mere size reduction, however, as the arm (in the male at least) appears to be specialized to act as a kind of support during mating, and as such is an analog of the clasper of a male shark, itself a highly modified fin.

Carnotaurus is usually pale green with horizontal yellow stripes and is adorned with surface armor, especially on the face and around the horns, that varies in pattern among individuals. Horns and ornamentation are found in both sexes. Carnotaurus lays between eight and ten eggs in a nest of rotting vegetation and earth. Both parents care for the young assiduously, teaching them how to hunt and track sauropod prey.

Habit and Habitat: These dinosaurs are usually found in small packs of two or three in lightly wooded to open country. They are often associated with herds of the medium-sized armored sauropod Saltasaurus, although they track and kill many other dinosaurs and, like most theropods, take smaller prey or carrion. Packs corner an individual prey animal, and each predator harrasses the hapless victim by ramming it repeatedly with its horns and bony head casque. Eventually the prey collapses, whereupon the predators slash at its head and throat.

Carnotaurus laying eggs; the nest will later be covered with more vegetation and guarded until the eggs hatch

An adult leads four young chicks on a mock hunting trip

	Triassic		Jurassic		Cretaceous	
	245m	208m		146m		65m
Saurischia						
Theropods						
Abelisaurs						

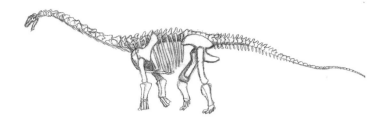

SALTASAURUS

Description: Medium-sized armored sauropod
Length: 33–43 ft (10–13m) nose to tail

Distinguishing Features: Heavy armor is the principal distinguishing feature of this dinosaur, one of the last sauropods. The ash-gray to reddish skin is punctuated by large, bony scutes on all parts of the body except the undersurface. A double line of smaller scutes runs along the spine, each capping a vertebral neural spine. The lateral side of the face is protected by a mask of thick bone, although the top of the face is left exposed, revealing large, inflatable nasal sacs. These creatures are gregarious, living in mixed herds of up to 80 animals with a social structure based strongly on clan relationships. They tend to seek mates in the same clan but with the most distant relationship to themselves —almost certainly reacting to pheromonal signals detected by a keen sense of smell. Females scrape craterlike depressions in sandy soil in which they lay between 40 and 60 eggs—a very large number. Fewer than 10 hatch, and it is thought that the remaining, unhatched eggs provide a ready source of nutritious food for the blind, relatively underdeveloped hatchlings.

Habit and Habitat: Found in habitats ranging from open, sparse coniferous woodland to semiarid scrubland, Saltasaurus lives on a rather poor diet of cones and tree needles, although these sauropods also dig for roots and scavenge for carrion. Their preference for open country exposes them to predation, and so the early warning provided by their sense of smell is vital. Even a distant threat from a large theropod such as Carnotaurus or Aucasaurus prompts these animals to form a defensive formation around their nesting ground. Aucasaurus has learned to exploit this defensive posture and while the adult predators create a distraction, the zebra-headed juveniles sneak in and steal eggs and babies.

Two male saltasaurus squaring off, lashing their tails threateningly

Profile of juvenile saltasaurus head

saltasaurus foot

Profile of adult saltasaurus head, stripping vegetation. Note armor around the eye and inflated nasal pouch

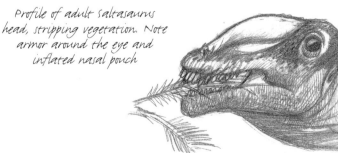

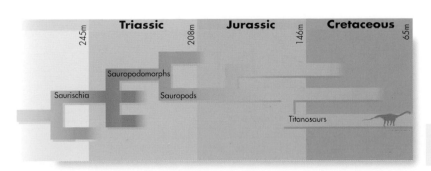

Triassic | Jurassic | Cretaceous

245m 208m 146m 65m

Sauropodomorphs

Saurischia

Sauropods

Titanosaurs

Overleaf: Juvenile Aucasaurus dodge the defensive ring to gain access to the saltasaurus nests

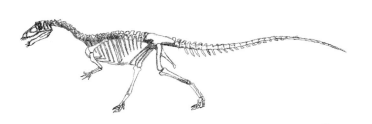

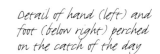

Masiakasaurus
diving after fish

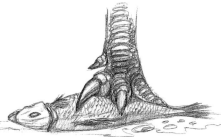

MASIAKASAURUS

Description: Small theropod
Length: 5–6.5 ft (1.5–2m) nose to tail

Distinguishing Features: This gray to light green dinosaur has a thin pelage of brownish filoplumage. Its teeth are its most remarkable feature—unusually for theropods, they vary greatly in size and are very large and procumbent toward the front and tip of the snout. Masiakasaurus is crepuscular, very shy, and among the most rarely seen of all dinosaurs. It has been observed alive on fewer than ten occasions: Always from a distance, late in the evening, perched on boulders along the banks of a fast-flowing stream, or amid whitewater rapids. Examination of the stomach contents of carcasses, together with the unusual dentition, shows that it lives almost wholly on fish. A capable swimmer, it dives underwater at the merest hint of a disturbance, and otherwise confines itself to dense woodland. Nothing is known about its social or sex life. Eerie, vixenlike vocalizations that one naturalist likened (perhaps somewhat fancifully) to "an electric guitar screaming in pain" have been attributed to it.

Habit and Habitat: This dinosaur is a member of the strange fauna that evolved in isolation In Late Cretaceous Madagascar. Masiakasaurus belongs to a group of theropods called abelisaurids, which are known only from Gondwana. Another abelisaurid, the large theropod Majungatholus, was also confined to Madagascar along with several unusual birds, such as Rahonavis. The absence of hadrosaurs and other ornithopods—typical of other parts of the world at the time—left opportunities for other herbivores, including a range of uniquely herbivorous crocodiles, and large titanosaurid sauropods such as Rapetosaurus.

Detail of hand (left) and foot (below right) perched on the catch of the day

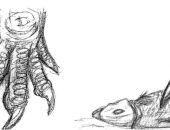

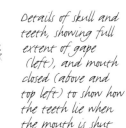

Details of skull and teeth, showing full extent of gape (left), and mouth closed (above and top left) to show how the teeth lie when the mouth is shut

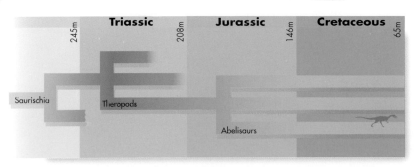

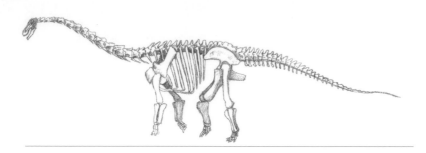

Detail of Rapetosaurus
head in side view

RAPETOSAURUS

Description: Medium-sized sauropod
Length: 40–55 ft (12–17m) nose to tail

Distinguishing Features: This medium-sized, lightly built sauropod resembles Diplodocus in its build and long, narrow skull but is in fact a titanosaur, the last and most successful group of sauropods. It is reddish with lighter, pink patches on the underparts, flanks, and upper limbs. The neck and upper body are protected by purplish-black bony scutes. Atypically for a sauropod of its era, it does not live in large herds but is found either alone or in groups of two or three. Mating occurs in the spring, after which females lay clutches of up to 20 eggs, only a few of which hatch. The females do not feed during the incubation period of 4 to 5 weeks. After hatching, the young develop very quickly and soon accompany their mother on foraging trips.

Habit and Habitat: This animal is found in many settings, from dense jungles through gallery forests and semiwooded parkland and swamps. It browses on low-growing plants and tender shoots of trees and spends much of its time wallowing in the mud of mangrove swamps. Titanosaurs represent the final flourish of sauropod evolution and almost all Late Cretaceous sauropods belong to this group. Elsewhere in the world they compete with ceratopsians and hadrosaurs, but on isolated Madagascar Rapetosaurus has few herbivore rivals. These circumstances have allowed it to become an effective all-around browser and grazer. Its main enemy is the large abelisaurid theropod Majungatholus, which it avoids by hiding in dense woodland or in swamps.

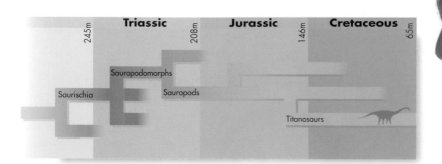

Triassic — 245m — 208m
Jurassic — 146m
Cretaceous — 65m

Sauropodomorphs

Saurischia

Sauropods

Titanosaurs

A swimming Rapetosaurus provides transport for a flock of the primitive bird, Rahonavis

Detail of the bony scutes found on the back of Rapetosaurus

A group of sauropods is surprised by the crocodile Majungasuchus, which bites at the snout of a sauropod venturing too close to the waterline

A Rapetosaurus is attacked by the abelisaurid theropod, Majungatholus

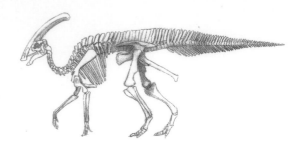

A herd of Charonosaurus on the move

CHARONOSAURUS

Description: Large crested hadrosaur
Length: 30–43 ft (9–13m) nose to tail

Distinguishing Features: This large hadrosaur—one of the last and largest of its kind—is a dull greenish color, with red patches on the snout. The head crest is black, while the flag-like flap of skin suspended from it, joining the neck, is red with a black border. The largest individuals are longer than *Tyrannosaurus rex* and half as large again as Parasaurolophus, a close North American relative. Only Shantungosaurus, another Asian hadrosaur, is larger. The enormous crest is the unmistakable feature of this animal, and resembles that of Parasaurolophus, but it is both absolutely and relatively longer—more than 6 ft (2m)—and fatter. In addition to the crest, the animal has high neural spines on its vertebrae, giving it an even larger and "high sided" appearance. It has particularly long forelimbs, suiting it for a largely quadrupedal habit, although it can walk and run on its hind legs if it needs to. As with all hadrosaurs, Charonosaurus has a busy social life, living in mixed herds that can number more than 500 animals of both sexes and all ages. Mating is promiscuous, and males engage in noisy displays for access to mates. Females harbor sperm in an internal store, or spermotheca: The clutches of 10–20 eggs may have varied paternity.

Habit and Habitat: Charonosaurus is most often found in dense, dark coniferous forest and overgrown, choked swampland, wherein its complex tooth batteries, of typical hadrosaur vintage, make short work of fibrous conifer needles and cones. Poor eyesight is more than compensated for by excellent hearing, which matches its spectacular vocalizations, "songs" of great subtlety and immense power. The sonority of these songs owes a great deal to the nasal crest, which contains a very large, hollow sinus continuous with the nasal passages. The most notable feature of Charonosaurus songs—or "ragas"—is the species' ten-octave range, from piccolo, almost bat-like twitterings to earth-shaking sub-bass. Detailed recordings suggest that these songs are partly inherited, partly learned, and are clan-specific.

A male Charonosaurus (right) inflates its throat pouch in display to a female

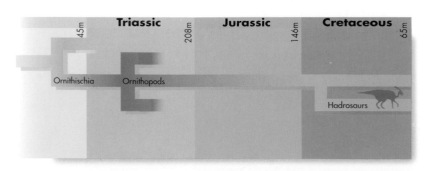

	Triassic		Jurassic		Cretaceous	
245m		208m		146m		65m
Ornithischia	Ornithopods					
					Hadrosaurs	

Head of contempory hadrosaur Corythosaurus

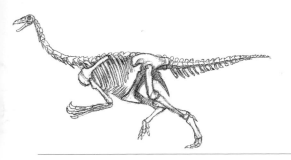

DEINOCHEIRUS

Description: Large herbivorous theropod
Length: 23–40 ft (7–12m) nose to tail

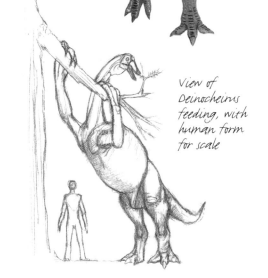

View of Deinocheirus feeding, with human form for scale

Distinguishing Features: Impossible to mistake for any other dinosaur except possibly a therizinosaur, this enormous creature, a gigantic relative of Ornithomimus and Gallimimus, combines the attributes of the ostrich, giraffe, and sloth. Blue-gray in color with distinctive red markings on its back, neck, and throat, it has a broad chest and abdomen, a relatively short tail, and powerful legs. Its most unusual features are its arms, which seem out of proportion to the rest of the body. These bear fearsome claws that contrast with its tiny head and toothless jaws.

The animals live in small family groups, and the meeting of two clans is often very noisy. Males display in leks before prospective mates, making threatening gestures with their arms (see main illustration), accompanied by a strident, percussive rattle as the claws on each hand are violently struck together. The noise, which can be heard from a great distance, has been described by one naturalist as the sound of "castanets from hell." Males and females build nests beneath large trees and take turns incubating the eggs. The hatchlings have arms and legs of equal length, as well as long prehensile tails, and are expert climbers almost as soon as they hatch, learning to move in slothlike-fashion beneath branches, feeding on young shoots, and stealing bird and pterosaur eggs.

Habit and Habitat: These creatures browse in woodland varying from dense forests to semiwooded parkland, using their arms to pull branches down to within reach of their jaws. This activity causes immense damage, yet it is thought that regular disturbances by large herbivores contribute to healthy forest succession. They eat virtually nothing but leaves, which are slowly broken down in the crop and stomach by symbiotic bacteria. Although apparently ungainly, Deinocheirus has no enemies. Its large size and immense claws ensure that even large theropods and ceratopsians usually keep well out of swiping range.

Deinocheirus in front view, showing arms at rest

Arm of Deinocheirus (left) compared with that of Therizinosaurus, another herbivorous theropod with long arms. Compare the long, grappling-hook-like arm of Deinocheirus with the more birdlike pose of the arm in Therizinosaurus

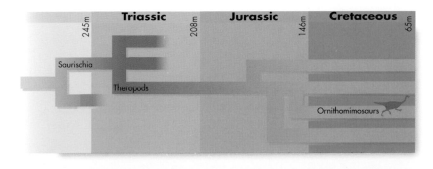

Deinocheirus stripping vegetation with its claws

	Triassic	Jurassic	Cretaceous

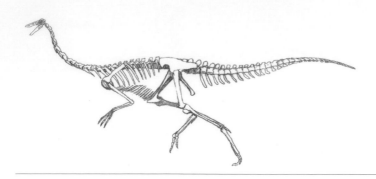

GALLIMIMUS

Description: Large ornithomimosaurid theropod
Length: 13–20 ft (4–6m) nose to tail

Distinguishing Features: Ornithomimosaurs are a diverse group of "ostrich-mimics:" light, fast-running theropods with long necks and tails, small heads, and large eyes. In some species, teeth have been replaced by horny beaks. Gallimimus is the largest ornithomimosaur, second only to the huge Deinocheirus. Typically colorful, Gallimimus is adorned with bold red and brown stripes, a distinctive, white, feathery mane, white feathering around the face, and a black cap of plumes on the top of the head. The beak is typically bright red, though this varies among different Gallimimus species. The dinosaur pictured here is *Gallimimus bullatus*, a gregarious animal found in flocks that can be up to 2000 strong, although this is exceptional. Males and females pair for life; they nest in large rookeries, laying clutches of six to eight small, blue eggs in sandy scrapes in the ground. Males and females take turns incubating the eggs. The chicks are covered with cream-colored down which is soon shed, and they can walk and run straight out of the egg.

Habit and Habitat: Gallimimus flocks live in lightly wooded parkland to semidesert scrub. They are omnivores, taking lizards, small mammals and birds, amphibians, and carrion. In the spring, they congregate around shallow, ephemeral lakes, wading offshore to catch seasonally abundant shrimp and small fishes. The main enemy of Gallimimus is the tyrannosaur Tarbosaurus, although Gallimimus can easily outrun it unless ambushed. Although not the fastest dinosaur, Gallimimus can sprint at 30–45 miles (50–70km) per hour.

Detail of Gallimimus
hand and foot

Gallimimus pulling down
branches to feed on shoots

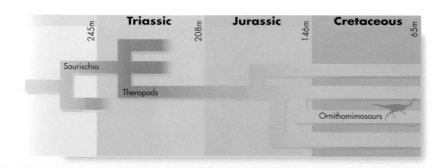

	Triassic		Jurassic		Cretaceous	
	245m		208m		146m	65m
Saurischia						
	Theropods					
					Ornithomimosaurs	

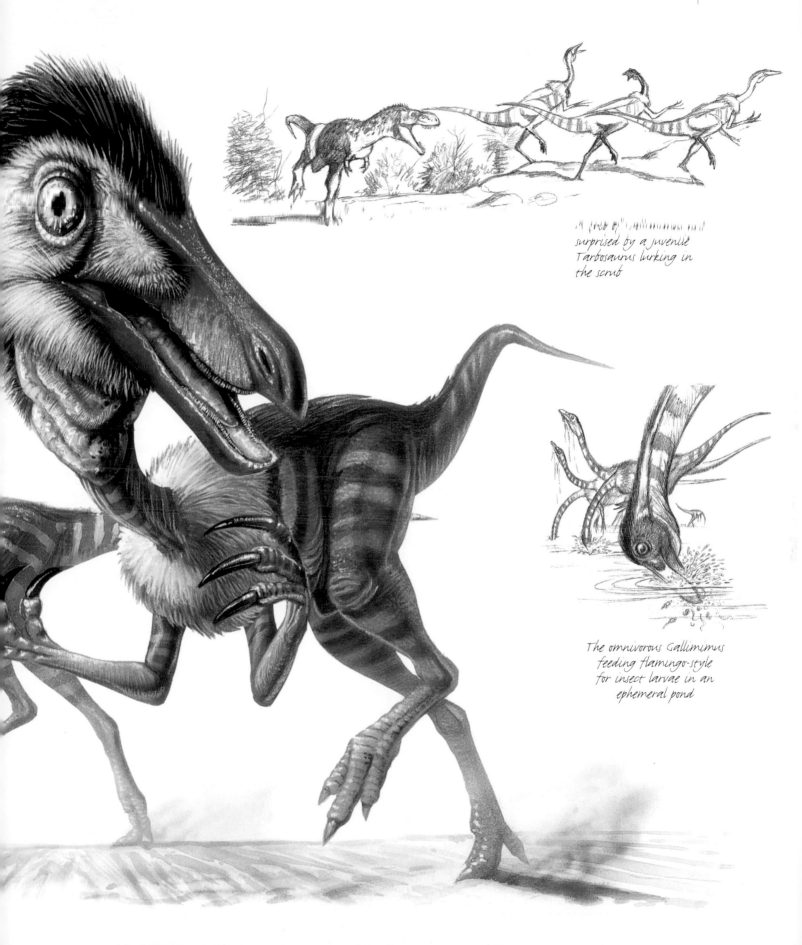

... ꞏꞏꞏ ꞏꞏ ... surprised by a juvenile
Tarbosaurus lurking in
the scrub

The omnivorous Gallimimus
feeding flamingo-style
for insect larvae in an
ephemeral pond

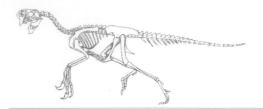

OVIRAPTOR

Description: Small oviraptorosaurian theropod
Length: 5–8 ft (1.5–2.5m) nose to tail

Distinguishing Features: This rangy, lightly built theropod is easily distinguished by the peculiar shape of its head—short and deep, with jaws modified into a sharp, horny beak—a huge gape, a bright blue face, and a prominent red crest. It is one of several rather similar, closely related dinosaurs—the main illustration shows *Oviraptor philoceratops*. This animal is thickly covered in a yellow to gray pelage, and the arms and tail carry a fringe of long, dun-colored feathers. The arms are long, with very long, clawed fingers. Like many theropods these creatures have an elaborate social life in which males display loudly and collectively to an audience of females; males and females form nesting pairs, yet both males and females seek to mate outside the pair. The eggs are incubated mostly—though not exclusively—by the female, and the clutch of up to seven or eight altricial chicks are raised by the efforts of both parents who feed them a mush of regurgitated, semidigested carrion.

Habit and Habitat: *Oviraptor philoceratops* is found in open, semidesert scrub and associates with herds of the ceratopsian Protoceratops in the way that zebras associate with wildebeests in the modern Serengeti. The nests of both animals are found close together. The fast-moving, sensitive, intelligent Oviraptor provides early warning of attacks from other theropods such as Tarbosaurus and Velociraptor. Oviraptor takes small prey such as the dinosaur Shuvuuia, lizards, snakes, and birds, but it is particularly fond of small mammals. This aspect of its diet suggests that this small, unusual dinosaur—active in the evenings and early mornings like its prey—serves to keep potentially egg-stealing vermin away from the communal nesting site.

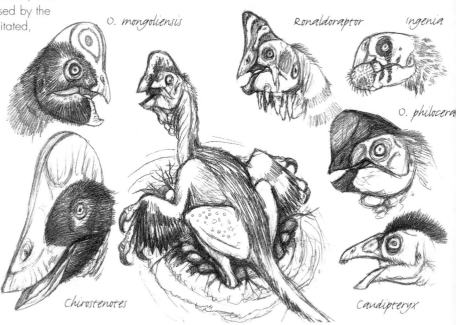

O. mongoliensis Ronaldoraptor Ingenia

O. philocera

Chirostenotes Caudipteryx

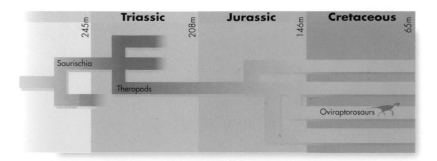

Triassic | Jurassic | Cretaceous

245m | 208m | 146m | 65m

Saurischia

Theropods

Oviraptorosaurs

Oviraptor philoceratops on its nest, surrounded by heads of various oviraptorosaurs for comparison. Clockwise from top: Ronaldoraptor; Ingenia (grasping a spiny fruit in its jaws); O. philoceratops in close-up; Caudipteryx; Chirostenotes; and O. mongoliensis

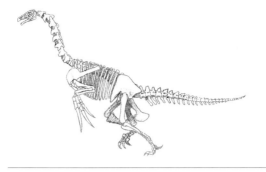

A therizinosaur uses its beak and long claws to strip bark from tree trunks, exposing the sapwood, fungi, and insects

Detail showing how the animal uses its snout to dig into the wood

THERIZINOSAURUS

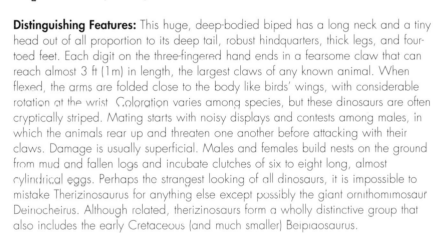

Description: Large herbivorous theropod
Length: 30–43 ft (9–13m) nose to tail

Distinguishing Features: This huge, deep-bodied biped has a long neck and a tiny head out of all proportion to its deep tail, robust hindquarters, thick legs, and four-toed feet. Each digit on the three-fingered hand ends in a fearsome claw that can reach almost 3 ft (1m) in length, the largest claws of any known animal. When flexed, the arms are folded close to the body like birds' wings, with considerable rotation at the wrist. Coloration varies among species, but these dinosaurs are often cryptically striped. Mating starts with noisy displays and contests among males, in which the animals rear up and threaten one another before attacking with their claws. Damage is usually superficial. Males and females build nests on the ground from mud and fallen logs and incubate clutches of six to eight long, almost cylindrical eggs. Perhaps the strangest looking of all dinosaurs, it is impossible to mistake Therizinosaurus for anything else except possibly the giant ornithomimosaur Deinocheirus. Although related, therizinosaurs form a wholly distinctive group that also includes the early Cretaceous (and much smaller) Beipiaosaurus.

Habit and Habitat: This animal lives in a variety of habitats, but typically in rather degraded swamp forests. Its claws are used to threaten the few predators it encounters, such as the Tarbosaurus shown in the main illustration. Their main purpose, however, is to bring high branches within reach of the jaws, as well as to strip bark from trunks. Therizinosaurus consumes leaves, rotted wood, leaf mold, fungi, insects, worms, and general woodland detritus, which it digests with the help of an army of bacteria, fungi, and other gut symbionts. It is thought that the claws are used to break open termitaries. This activity has never been observed, although therizinosaurs have formed symbioses with cellulose-digesting worms that live in their gizzards and intestines. Close examination shows that these worms are in fact highly degenerate termites of a kind known only from therizinosaur guts. The volume of methane produced by cellulose digestion on this scale is considerable, although tales describing therizinosaurs struck by lightning and exploding into fireballs of blue flame are probably apocryphal.

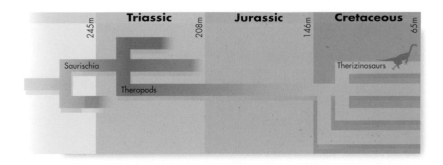

	Triassic		Jurassic		Cretaceous	
245m		208m		146m		65m
Saurischia					Therizinosaurs	
	Theropods					

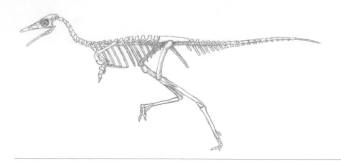

SHUVUUIA

Description: Small, feathered theropod
Length: 12–24 in (30–60cm) nose to tail

Distinguishing Features: This lightly built dinosaur has long, spindly legs that contrast with its short, winglike arms, each of which terminates in a single prominent claw. It has prominent black and white plumage, with a black head crest and tail plumes. The distinctly beaklike snout is yellow to brown and bears small teeth near the front of the jaws. It is very hard to distinguish Shuvuuia from its close relative Mononykus. At first sight, glimpsed from afar across the plains, it looks like a medium-sized ground bird, but it is not in fact closely related to birds. Males and females tend to pair for life and make nests on the high shoulders of old termite mounds or rock outcrops, from which they defend large territories against other breeding pairs.

Habit and Habitat: Shuvuuia lives in semiarid savannah to desert margins where it is a specialist insectivore. It breaks into termite nests with its feet and powerful fore claws, inciting the residents to attack. It also uses its fore claws to strip bark from trees to excavate the insects beneath (see Therizinosaurus). Protected from bites by thick plumage, the animal mops up insects with its very long, barbed tongue. It can retain a ball of half-digested insects in its gizzard for later feeding to its chicks. Its main enemies are small theropods such as Oviraptor and Velociraptor, against which its only defense is running at great speed and kicking dust into the face of its pursuer in the same way that a squid generates clouds of ink. The animal can use its short wings to generate enough thrust for running up the steep slopes of termite mounds.

Detail of shuvuuia eggs

A robust claw emanates from the single, enlarged digit on each stubby forelimb

Typical "tree-hugging" behavior, in which shuvuuia strips bark from a tree in search of insects

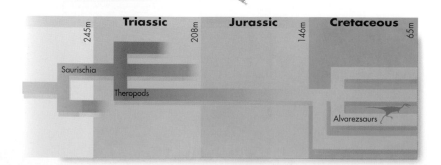

	Triassic	**Jurassic**	**Cretaceous**	
245m		208m	146m	65m

Saurischia

Theropods

Alvarezsaurs

A mating pair of shuvuuia

a disused termite mound
to make a nest

shuuuuia attempts to
maintain to pharyingress
predator, throwing up a
cloud of dust as it goes

PROTOCERATOPS

Description: Small ceratopsian
Length: 5–9 ft (1.5-3m) nose to tail

Distinguishing Features: Protoceratops has the size, shape, and temperament of a large pig, along with a barrel-shaped body, short, powerful legs, and a deep tail. As in all ceratopsians, the head is dominated by a prominent neck frill adorned with bony plates. The frill is more prominent in the male than in the female. Males also have nasal horns (in females the horns are small or absent) and facial tusks. Both sexes, however, have extremely powerful, parrotlike beaks. These animals are a uniform reddish brown. They are invariably found in large herds that may number 200 or 300. Closer inspection reveals that these herds are divided up into groups of females and subadults dominated by single, dominant males; bachelor males roam through the herd seeking opportunities to mate and sometimes challenge dominant males to vigorous head-butting contests. Like the related Zuniceratops from North America, Protoceratops tends to nest communally, producing large, craterlike depressions of mud and vegetation. The smaller nests of the theropod Oviraptor may be found wedged in between.

Protoceratops in front, lateral, and dorsal views

Habit and Habitat: Protoceratops are generally found in dry, open country where they feed on tough vegetation, roots, and occasional carrion. The large herds and concentrations of eggs attract a wide range of unwelcome visitors, the smaller of which are observed near the flocks of Oviraptor invariably found associated with Protoceratops herds. Larger predators such as Tarbosaurus may be deterred by the phalanxlike communal defense of male Protoceratops working together, and small theropods such as Velociraptor can be overcome by a single, grimly determined male (see main illustration).

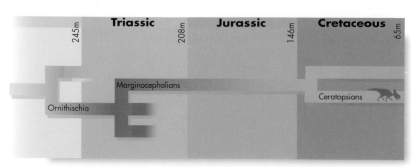

A male Protoceratops stands its ground against attackers

	Triassic	Jurassic	Cretaceous
245m		208m	146m	65m

Marginocephalians

Ceratopsians

Ornithischia

VELOCIRAPTOR

Description: Small theropod
Length: 6–8 ft (1.8–2.4m) nose to tail

Distinguishing Features: The animal has a reddish skin colored with black and white plumage on the head, neck, flanks, arms, and tail. The head is distinctively long and carried low, without the tendency toward deep jaws seen in many other theropods. Like Deinonychus, the second toe carries an enlarged, slashing claw. Unlike its relative, however, the sexes of Velociraptor closely resemble one another and they tend to look more like the crested males of Deinonychus rather than the bald females. Matings occur at the end of a ritualized, threatening dance in which the male runs a high risk of suffering mortal wounds. This behavioral trait imposes a terrific selective pressure on ferocity in both males and females. It may also explain why both sexes look like males: The most successful animals hatch from eggs loaded with male hormones, and so tend to resemble males whatever their sex (determined chromosomally by the ZW system used in birds).

Habit and Habitat: Velociraptor is found in a wide range of habitats, from swamp forests to dry, open country. It subsists on the young and eggs of Protoceratops and Therizinosaurus, as well as smaller dinosaurs such as Shuvuuia and Oviraptor, and a range of smaller prey. Despite its popular depiction as an intelligent pack-hunting animal, any such instincts in these creatures have been submerged by an inherent viciousness. The packs are not social groups, but simply collections of individuals drawn by a common lure. After making a kill, the animals typically fight one another to the death for possession of the spoils. Indeed, Velociraptor attacks virtually anything that moves with an unprovoked recklessness: Lone individuals have taken on gigantic adult Therizinosaurus, and even Tarbosaurus, with lethal consequences for the smaller dinosaur. Needless to say, this animal should never be approached.

A pair of Velociraptor in the threatening pre-mating dance

	Triassic	Jurassic	Cretaceous
	245m	208m	146m	65m

Saurischia

Dromaeosaurs

Theropods

Hand, showing the wrist action reminiscent of a bird's wing-folding mechanism. This is found in many theropods, especially dromaeosaurs and therizinosaurs

Foot, showing the extent to which the enlarged second digit can be lifted clear from the ground

Head in front and side view

Velociraptor in dorsal, side, and front views. Note the extremely long tail, capable of being bent straight upward—a trait that is an important part of the animal's threat display

GLOSSARY

Abelisaurs
A group of theropods typically found in the southern continents, e.g. Masiakasaurus and Majungatholus.

Ammonites
A group of mollusks related to modern squid, made distinctive by their heavy and often highly ornamented, coiled shells, which grew throughout life.

Ankylosaurs
A group of heavily armored, herbivorous dinosaurs found worldwide in the Cretaceous period e.g. Minmi and Edmontonia.

Archosaurs
The group of reptiles that includes dinosaurs, pterosaurs, crocodiles, birds, and some other extinct forms, but not turtles, snakes, or lizards.

Belemnites
A group of extinct, squidlike mollusks. Unlike ammonites, their shells were internal.

Brachiosaurids
A group of giant sauropods typified by the Jurassic Brachiosaurus.

Carnivore
Strictly speaking, any mammal in the Order Carnivora. Used more loosely, the term refers to any creature adapted for a diet consisting chiefly of live meat.

Ceratopsians
A group of mostly horned, mostly quadrupedal, ornithischian dinosaurs whose heyday was in the Late Cretaceous and included forms such as Psittacosaurus, Protoceratops, Zuniceratops, and Triceratops.

Cladistics
A scheme of classifying species according to their evolutionary relationships, without reference to general similarity or relative position in geological time.

Coelacanths
A group of archaic fish, distantly related to land vertebrates, believed to have become extinct in the Mid-Cretaceous until living forms were discovered in the Indian Ocean in the twentieth century. The classic "living fossils."

Continental drift
The process whereby landmasses move across the face of the Earth, driven by the forces of plate tectonics.

Crustaceans
A large group of arthropods (jointed-legged animals), mainly aquatic, that includes crabs, lobsters, shrimp, barnacles, and many other forms.

Dromaeosaurs
A group of small, bipedal theropods from the Cretaceous period, among the closest dinosaur relatives of birds. They include Deinonychus, Velociraptor, Sinornithosaurus, Microraptor, and others.

Fossilization
The process whereby organic remains are preserved in rock, usually by the slow replacement of hard tissues such as bone by minerals carried in groundwater.

Hadrosaurs
A group of specialist, herbivorous dinosaurs, often with very distinctive head ornamentation, which were very successful in the Cretaceous period. Examples include Charonosaurus, Lambeosaurus, Corythosaurus, and Parasaurolophus.

Herbivore
Any animal chiefly adapted for living on live plant material.

Ice Ages
Cold snaps in a global climate that lead to a long-term increase in polar icecaps and glaciers elsewhere. The term conventionally refers to the several fluctuations of the past 2 million years, but there were Ice Ages in the Permian period and at least two intervals before 500 million years ago in which the Earth may have been completely covered with ice.

Ichthyosaurs
A group of Mesozoic reptiles highly adapted to marine life and which looked very similar to dolphins. They were not closely related to dinosaurs.

Insectivore
Any animal chiefly adapted for living on live insects or other small prey.

Jacobson's Organ
Also known as the vomeronasal organ, Jacobson's Organ is a patch of sensory cells in the roof of the mouth of many animals, including humans, but most developed in reptiles such as snakes. It is used as an accessory nose, to detect various kinds of scent.

Land bridges
Before the theory of continental drift, similarities in fauna and flora between distant continents were explained by the proposal of causeway-like "land bridges" between continents that had since sunk or been eroded.

Leks

These are "dance floors" in which members of one sex of a species, usually the male, will display before an audience of the other sex.

Mesozoic era

The interval between 245 and 65 million years ago during which time the dinosaurs lived. The Mesozoic is divided into the Triassic, Jurassic, and Cretaceous periods.

Ornithopods

A group of mainly bipedal, herbivorous, Early and Mid-Cretaceous ornithischian dinosaurs e.g. Iguanodon, Tenontosaurus, and Ouranosaurus, largely superseded in the Late Cretaceous by the more specialized hadrosaurs.

Pachycephalosaurs

A group of mainly bipedal, herbivorous, typically Late Cretaceous ornithischian dinosaurs, distinguished by very thick, domed, and armored skulls. Examples include Pachycephalosaurus, Homalocephale, and Stygimoloch.

Paleontology

The study of fossils.

Parasite

Any organism adapted for living as a freeloader, completely at the expense of another. Examples include tapeworms and flukes in the gut, and the microscopic malarial parasite of the blood.

Parthenogen

An animal that can reproduce clonally, that is, without fertilization. It follows that all parthenogens are females. Many invertebrates are habitual parthenogens, but some species of amphibian and reptile may be parthenogenetic on occasion. No known mammal or bird is parthenogenetic.

Plesiosaurs

A group of Mesozoic reptiles adapted for life in water, though not to the extreme extent of ichthyosaurs. They had long tails and necks, and a rounded body bearing two pairs of flippers. Examples include Plesiosaurus and Elasmosaurus.

Pliosaurs

More fearsome than the plesiosaurs were the short-necked pliosaurs, which included some of the most fearsome predators ever to have evolved. Kronosaurus for example, was 40 feet (12m) long, almost 13 feet (4m) of which was skull.

Plumage

The description of the disposition and color of feathers in a bird or dinosaur.

Pterosaurs

A group of Mesozoic flying reptiles e.g. Rhamphorynchus, Pteranodon, Quetzalcoatlus, and Pterodactylus, closely related to dinosaurs. Some are thought to have had a hairlike integument, but none are known to have had feathers.

Sauropods

A group of herbivorous saurischian dinosaurs e.g. Brachiosaurus, Diplodocus, and Isanosaurus.

Stegosaurs

A group of armored, herbivorous ornithischian dinosaurs, distinguished by a series of large, dorsal plates e.g. Stegosaurus and Tuojiangosaurus. Largely superseded by ankylosaurs in the Late Cretaceous.

Subduction

The phenomenon in which material from one tectonic plate disappears beneath the edge of another, either in an ocean trench, or as part of a process of raising high mountains such as the Himalayas, caused by the subduction of India beneath Tibet. See also continental drift.

Therizinosaurs

The group of peculiar theropod dinosaurs, secondarily specialized for herbivory. Examples include Therizinosaurus and Beipiaosaurus.

Theropods

The large group of mainly carnivorous saurischian dinosaurs. Incredibly diverse, they include giant carnivores such as Tyrannosaurus and Allosaurus, as well as the bizarre therizinosaurs, ornithomimosaurs, and oviraptorosaurs, not to mention a host of small, often feathered forms including dromaeosaurs, troodontids, and birds.

Titanosaurs

A large and successful group of sauropods, particularly prominent in the Late Cretaceous. Examples include Rapetosaurus and Argentinosaurus, the largest known land animal.

Vertebrates

The large group of animals with backbones, including all fishes, amphibians, mammals, reptiles, and birds, living or extinct.

ZW system

The chromosomal system of sex determination found today in many birds in which males are homogametic (ZZ) and females heterogametic (ZW). Contrasts with mammalian "XY" system in which females are homogametic (XX) and males heterogametic (XY).

INDEX

Figures in italics indicate captions.

CREDITS

Quarto would like to thank and acknowledge the following for supplying pictures used in the book:

Page 15: Mick Ellison—AMNH
Page 24: Don Davis

After all the rude things Henry Gee has said about "Walking with Dinosaurs," he never imagined he'd be working on a similar project. He would like to thank Luis, the world of paleontology, and Fred and all the girls for inspiration. The Cranley is gone but not forgotten.

Luis Rey would like to especially thank:
Per Christiansen, Darren Naish, Scott Hartman, Nick Longrich, Marco Signore, Luciano Campanelli, Mickey Mortimer, Jaime Headden, Ken Carpenter, Tom Holtz, Midori Sugimoto, Charlie and Florence Magovern, John Hutchinson, Scott Sampson, David Lambert, Sandra Chapman, Robert Bakker, Mary Kirkaldy, Janet Smith, Dick Pierce, Mark Kaplowitz, Henry Gee, and my partner Carmen Naranjo for relentless help and inspiration.
Also my thanks go (among many!) to:
Eberhard "Dino" Frey, Eric Buffetaut, David Eberth, Don Brinkman, David Martill, David Unwin, Eric Buffetaut, Cristiano Dal Sasso, Mark Norell, Mick Ellison, David Peters, Alan Gishlick, Mike Skrepnick, Osamu Miyawaki, Mike Taylor, Greg Paul, Tracy Ford, John Lanzendorf, George Olshevsky and everybody at Quarto Publishing for their dedication and hard work.
To the memory of Mr. Maasaki Inoue and everyone that left us from 1999 to 2002.